Canoe *for* Change

A Journey Across Canada

GLENN GREEN AND CAROL VANDENENGEL

 FriesenPress

Suite 300 - 990 Fort St
Victoria, BC, V8V 3K2
Canada

www.friesenpress.com

ISBN
978-1-03-910301-6 (Hardcover)
978-1-03-910300-9 (Paperback)
978-1-03-910302-3 (eBook)

1. BIOGRAPHY & AUTOBIOGRAPHY, ADVENTURERS & EXPLORERS

Distributed to the trade by The Ingram Book Company

For Eric
wherever you may be

Table of Contents

Prelude   1

Introduction   3

Year One: Ottawa, Ontario to Sydney, Nova Scotia   11

     Lower Ottawa River   13

     Saint Lawrence River   17

     Madawaska, Saint John and Canaan Rivers   27

     Northern Northumberland Strait   35

     Southern Northumberland Strait   43

     Bras d'Or Lake   51

Year Two: Vancouver, British Columbia to Fort Frances, Ontario   59

     Pacific Ocean and Fraser River   61

     Rail Trails   67

     Rocky Mountains   73

     Oldman River   83

     South Saskatchewan River   85

     Saskatchewan River   117

     Lake Winnipeg   125

     Winnipeg River, Lake of the Woods and Rainy River   137

Year Three: Fort Frances to Kingston, Ontario   **145**

    Boundary Waters   147

    Northern Lake Superior   159

    Northeastern Lake Superior   167

    Eastern Lake Superior   175

    Lake Huron   183

    French and Mattawa Rivers   187

    Upper Ottawa River   195

    Rideau Canal   199

Epilogue   **203**

Acknowledgements   **207**

References   **211**

About the Authors   **215**

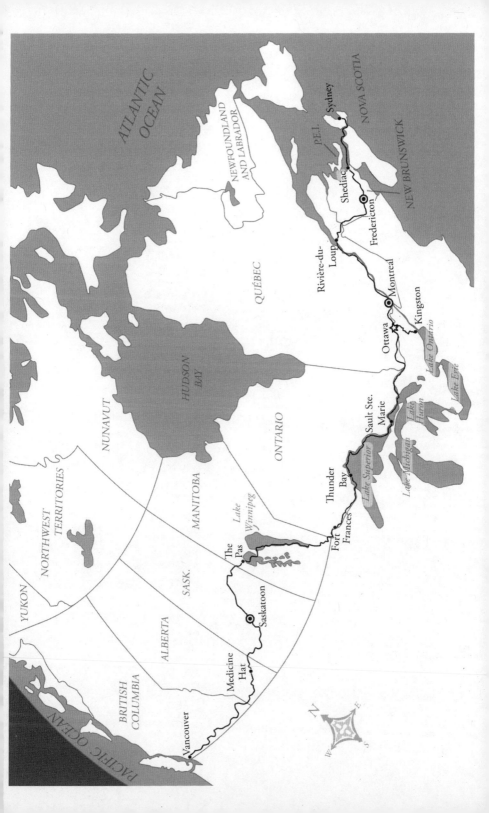

Prelude

Many days running a non-profit organization feels like pulling a canoe over gravel—every step is a challenge.

Once in a while, you find yourself running down river on a strong current, enjoying the pleasure of the trip.

Carol and Glenn were that strong current.

We met for coffee at a café on the shores of Lake Ontario in Kingston so I could hear about their dream. Carol and Glenn were just about to enter retirement. They wanted to canoe across Canada and to raise awareness of the need for healthy food among all Canadians. I had the pleasure of shepherding Loving Spoonful, the Kingston-based good food organization, as the Executive Director. Our shared commitment to increasing food security was the link that created a great partnership.

Carol and Glenn's lifelong dedication to healthy food was obvious in their energy and an overall glow. I had no doubt that they could paddle long-distances, but canoeing across Canada? How does one do that? The rivers are not connected—there would be miles of portages, the ferocity of the Great Lakes, and the great Rocky Mountains in their way. I listened and caught their enthusiasm, then went home to tell my family about a wild dream that a couple had—to canoe across Canada. I had many doubts. I doubted:

1. Whether anyone could have that much tenacity.
2. Whether a marriage could survive three months in a canoe.
3. Whether Canada could be canoed.

Answers

1. Yes. They can—and they do!
2. Yes—three years, in fact!

3. Yes—canoed, portaged, hiked, and weathered.

Nothing was stopping Carol and Glenn.

At Loving Spoonful, we tracked their travels on a huge map of Canada, following their SPOT tracker. We celebrated the tens of thousands of dollars they raised that we put right back into community programs for healthy food. Programs at Loving Spoonful included school gardens with full curriculums, healthy cooking with folks living with long-term mental illness, people experiencing poverty and the attendant food insecurity, Indigenous food programming, fresh food access, and so much more. Connect at www.lovingspoonful.org.

Carol and Glenn raised awareness and unexpected funds for Loving Spoonful, shared their hearts and energy, and constantly inspired all of us who watched their trips.

Three years later, they had overcome winds, waves, rain, mountains, and snow. They paused to address a life transition, then, as you will hear, resumed their trek. Glenn and Carol were unstoppable, unflappable, undeterred, and seemingly continuously joyous.

Three years after they left, they paddled into the docks at Kingston. The celebration was huge and heartfelt.

Enjoy this virtual trip with Carol and Glenn, dear Reader. Imagine that you are with them, winded on the shores of Lake Winnipeg, racing huge waves near the Atlantic, and climbing over the Rockies. You'll be cold, wet, sore—and inspired.

On behalf of both Loving Spoonful and personally, I am so grateful for Glenn and Carol's determination and generosity! We would have loved you whether you paddled one kilometre or eight thousand kilometres. Your eight thousand plus kilometres and years of dedication are like a beautiful day on calm waters: a sheer pleasure—and we get to share the trip!

Mara Shaw
November 2020

Introduction

Canada spans six time zones, reaches three oceans and has the longest coastline in the world. With over two million lakes and rivers, Canada is a paddler's paradise. We both enjoy the great outdoors and use the canoe as a means of finding tranquillity in nature. There is no better way to see this beautiful country than by canoe, a vessel of creation that has become a favourite Canadian pastime.

We had both had busy office careers which demanded much of our time. Glenn worked as an auto appraiser in the insurance industry; Carol as an office manager with a service-oriented landscape firm. This, after working for a not-for-profit charitable organization for many years. Over the past ten years, we moulded our vacations to fit our working schedules. We arranged backpacking treks to numerous continents around the world. We hiked in the Amazon, trekked through the Guatemalan Cuchumatanes Mountains and along the coast of Portugal. We went snorkelling in the barrier reefs of Thailand, sailed on a catamaran in the Belize Sea and camped at the bottom of the Grand Canyon. After a strenuous hike in the Mojave Desert, we were reminded of how fortunate we are to live in a country that has more water than any other nation in the world. It would be far easier to paddle on water rather than hike across barren deserts while worrying about when, and if, we would find our next water source. As much as we enjoyed travelling to other lands, we were inescapably drawn to those that took us on wilderness excursions in our homeland, Canada. For us, there is nothing more soothing to recharge our mind, body and soul than dipping our paddles into the calm waters of a northern lake to enjoy the peace and quiet of early morning mist on the water, or to watch a beautiful sunset cast its shadows on the forested land. Past experiences revealed a place of discovery in the forest and divulged to us a place of wonder and satisfaction of being on the water. We are all instinctively closely connected to nature, whether we realize it or not. We have been for thousands of years. Living close to the earth, we can enjoy each and every moment in the present time. We felt the need to do more of what we wanted but on a grander scale.

The time had come to recharge and to refocus on an exciting new chapter of our lives. Both of us turned sixty in March of 2017. Empty-nesters, our children, along with their life partners, live in different cities and are finding their own place in the world. While in good health, we had no time like the present to take on a new adventure. We notified our employers that we were retiring, paid off our debts, and decided to paddle across Canada. Happy and thrilled to embark on a canoe trip exploring our own country, we wished to complete in our retirement, a trek from the west coast to the east coast. We commenced planning our trip and, seven months later, it came to fruition.

We met many extraordinary Canadians on our journey, but one particular elderly gentleman confirmed that we were not alone in our convictions. Upon exiting Lake Winnipeg, we had the pleasure of meeting Ed. He told us that he had immigrated to Canada from New Zealand after he had served several years with the navy. Ed told us that in the early seventies he had had the opportunity to be flown to Nunavut with the Royal Canadian Mountain Police. Along with his canoe, he was deposited on an unknown river, the name of which he had long forgotten. From Nunavut, Ed paddled all the way south to the Amazon! He did remember he lived off fish and whatever he managed to catch, no matter how untasty. He rigged up a sail to harness the wind and was held up for three days in a northern tundra while he waited for a herd of migrating caribou to pass. We were humbled as we listened to Ed speak with genuine modesty about his incredible journey. Now old and frail, Ed was reliving his experience through us as he spoke of his trip that he had taken when he was young and healthy. When we asked him what made him decide to embark on such a grand adventure, he said, "Well, I never did really have a reason. I just did it because I wanted to." Much like Ed's, our reason to paddle across Canada merely is the same. There are others like Ed who have had much more incredible journeys than we have had. Together, we share a common bond of having our reasons for choosing what we feel the need to do. For us, it was to quench an insatiable thirst. Being on the water connects us to the natural world on a deeper level. It gives us food for our spirits and stillness for our minds. We cannot get enough!

Canada was celebrating its one hundred and fiftieth anniversary of Confederation. We decided to start at the nation's capital and head east. We would complete that portion the first year, return home then travel west to initiate the voyage back to the nation's capital to finish. Rivers and lakes of Canada were the original highways of Indigenous peoples and then of European exploration thereafter. We were to travel on this complicated system; the arteries and capillary network of water that stretched across the country used by so many before us. We were apprehensive. Extensive waterways like Lake Superior and the Atlantic Ocean awaited us. There were also kilometres of portaging connecting the waterways, with our being exposed to the elements for many weeks on end and living in a tent. We would not have the assistance of a support vehicle following us. Our gear would enable us to camp along the shoreline. Food would be packed, enough to sustain us for many weeks.

We set about planning. The preparation to canoe across Canada was an adventure in itself. We spent hours formulating a route connecting the waterways across Canada. From west to east the majority of the watersheds flowed in our favour. Where the water flowed against us, as was the case in British Columbia's Rocky Mountain range, a solution was needed to circumvent a distance of eight hundred kilometres. Besides preparing route logistics, we purchased expedition equipment. We acquired a suitable canoe—one that could withstand wind and waves for big-water travel on oceans, but that could manoeuvre intricately on meandering rivers. It also had to be light enough to portage. We purchased a sturdy, freestanding tent along with waterproof bags. We upgraded our food dehydrator to a larger unit which enabled us to prepare large quantities of what was required to sustain us. The list went on and on—with the expense going up and up.

Evaluating the route, we estimated that eighty per cent of the time we would not have access to an urban area to restock needed supplies. Careful menu planning was required to ensure we would not run out of food, particularly if hundreds of kilometres away from any community. Because we had to carry enough food for roughly eight weeks at a time, weight was a consideration. The food needed to be lightweight, easy to prepare, and consume as little cooking fuel as possible. We were familiar with planning and preparing meals from past wilderness canoe camping trips. We adopted the same methods but on a larger scale. We spent many hours dehydrating fresh produce for the journey. The produce was vacuum-sealed to preserve freshness. The food was then gathered, boxed and delivered to us during our trip at different locations across Canada. Where we did have access to communities, we restocked our food staples and fuel and replaced broken equipment as required.

We agreed that if we were to do this retirement cruise, others should benefit. We singled out a local hometown charity, Loving Spoonful, which is one that matches our belief 'that if you eat well, you feel well; and if you feel well, you do well.' Approaching Executive Director Mara Shaw, and informing her of our plans, we saw that she was taken aback at the idea. Still, she embraced the thought to increase awareness for Loving Spoonful. Mara's passion for the organization was evident. Her drive was contagious. This reaffirmed that we had made the right decision in choosing Loving Spoonful. The staff work hard to ensure that everyone in Kingston has access to fresh, healthy food. This is a fundamental right that all Canadians should have. Unfortunately, this right lacks in many corners of our country. By teaching elementary students and adults numbered in the hundreds how to grow gardens, Loving Spoonful is undertaking the sustainable approach to combat food insecurity at a local level. Dedicated staff, along with volunteers teach budgeting and cooking skills with fresh, healthy food. Many are recruited to deliver fresh produce donated from grocery stores to shelters and community programs. Our mission, as we paddle across Canada, is to raise funds as well as share the knowledge of Loving Spoonful's sustainable food-secure organization. It is a charity that connects people with fresh food and culture—a role model for the rest of Canada to follow. The hope is that someday all Canadians will have the right to access fresh, healthy food.

On the matter of why we choose food security, Carol reflects, "*From my experience, food has made a huge difference in the quality of my life. In my late twenties, I contracted a serious infection. I was ill for several months, and, as a result, I was left with a compromised immune system. For several years I struggled with declining health. I visited numerous specialists with no result. By chance, I was referred to a doctor who treated a variety of complex, chronic, medical conditions—a medical doctor practising functional medicine. The doctor placed me on a 'back to basics' diet that focused on eating a high intake of fruits, vegetables, legumes, whole grains and monounsaturated fats, a diet that eliminated foods such as refined sugar and processed food. I was put on a diet plan for two weeks to eliminate many of the foods I was used to eating. Some of these foods were not necessarily unhealthy, but the wrong food for me. After two weeks, the difference in my health was incredible. Not only did I change my diet, but I put my young family, as well, on a revised eating plan, one that we try with some modifications to maintain to this day. The transformation was so complete that I became an advocate for healthy eating—another reason why we chose 'food security' to be our cause. Glenn and I know firsthand that fresh, healthy food does indeed make a huge difference in our quality of life. We are truly fortunate to have a local agency, Loving Spoonful, which aligns with our beliefs.*"

Our cause is termed 'Canoe for Change'. Our friend Debra suggested this name. During the initial planning stages of our trip, Carol met with Debra at a local Kingston coffee shop. Once she learned that Carol needed a name for our journey, Debra came up with 'Canoe for Change'. She then proceeded to poll everyone in the coffee shop as to what they thought of the idea. Debra stated this would induce a change in how people view food security. By paddling across Canada, we might assist in that 'change'. After an enthusiastic response from patrons, 'Canoe for Change' was born!

Now, this evolved into a retirement cruise with purpose. Talks began about building a website, obtaining business cards and activating social network applications. All this electronic capability was new to us and a little overwhelming. We wanted to engage fellow Canadians toward food security awareness and felt this was a great way to do it. We did not pursue sponsorships from outdoor equipment sporting companies to promote their products. Our focus was on earning donations for the cause. We did, however, accept a gift for website creation to make this all happen.

Our average speed of paddling was to be at a slow pace of five kilometres an hour. This certainly would give us pause to appreciate the beauty of our country that we were to explore from the water. We anticipated paddling from the scenic coastline of the Pacific Ocean into some of the world's most dramatic mountain scenery. We would travel through the snow-capped towering peaks of the Canadian Rockies with cascading waterfalls, old-growth forests, glacial lakes and swollen rivers. Beyond the Rockies, we would paddle through the flat expanse of the sprawling prairies along meandering rivers which had been formed by ancient glacial melts to reveal dramatic cliffs of sandstone and the Badlands. The rivers would take us all the way to the largest freshwater delta in Canada as far north as Cumberland House, Saskatchewan. Eventually, this would lead us to the sandy shores of Lake Winnipeg: the unpredictable lake that many paddlers avoid at all costs. Then we would reach our home province of Ontario where we would have to navigate through the maze of hundreds of secluded lakes, rivers and windswept forests to the rugged shoreline of Lake Superior. Dramatic cliffs and crystal-clear waters of this largest freshwater lake in the world would bring us to some familiar territory past the towering cliffs of the French, Mattawa and Ottawa River systems. But first, the multiculturally diverse cities of the Saint Lawrence Valley would await us. Living in a tent and using a canoe for transportation would be a challenge on a whole different level. We would paddle the majestic Saint Lawrence

River, where billions of tons of tidal waters flow in and out of the river every six hours. Inland, we would go in search of the maritime rivers that would lead us to the salty ocean waters of the Northumberland Strait. Our little canoe would manoeuvre the ever-changing ocean waters along the shoreline to Cape Breton Island in Nova Scotia and thence to the vast expanse of the Atlantic Ocean.

Eventually—after many months of planning—our trip came together. So, after a hearty, emotional farewell from friends, family, and supporters, we dipped our paddles into the water. We anticipated a 'retirement cruise' of a lifetime. We set off from the beach on Petrie Island in Ottawa, Ontario, our nation's capital. Not knowing what lay ahead of us on this grand adventure, we launched our canoe with excitement and anticipation. At least our excursion would be better and more exciting than were we to sit at home with our slippers on.

Year One: Ottawa, Ontario to Sydney, Nova Scotia

August 12, 2017 to October 18, 2017

Lower Ottawa River
Ottawa, Ontario to Montreal, Québec

Dates: August 12 to 16

Route: Ottawa River; Saint Lawrence River

Retirement: how did ours begin? One paddle stroke at a time! As we left the beach on Petrie Island in Ottawa, we turned around to take one last glimpse of where we would end up at the end of our cross-Canada journey. We were determined to take this once-in-a-lifetime opportunity one day at a time to fully enjoy the journey. We did not consider this a race. A feeling of panic overcame us when we realized after the first day that we had only completed twelve kilometres—with eight thousand four hundred eighty-eight kilometres left to go.

Our destination was Sydney, Nova Scotia on the Atlantic Ocean, approximately one thousand eight hundred kilometres away. The goal was to paddle two hundred kilometres a week to reach Sydney by the end of October, the onset of winter. There would be many challenges to accomplish this goal. With autumn approaching we could expect shorter days, colder temperatures and windy conditions which are the after-effects of hurricane season common in the south Atlantic. Nevertheless, it was a calm sunny day in August. We had our sights set on paddling down the Ottawa River to the Saint Lawrence River. Then once in the Saint Lawrence, we would paddle to Rivière-du-Loup, Quebec, where we would go inland to portage towards the Saint John River, New Brunswick.

We had brought our seven-year-old granddaughter Esther along for the first few days of our journey. Paddling a canoe down a once-important voyager route was an excellent way to experience the rich history of her country. We paddled down the Ottawa River which forms the border between Ontario and Quebec. Esther pointed out that we could swim in Quebec, have lunch in Ontario and camp in Quebec all on the same day. She said, "It is nice to have choices." We fed her curiosity by taking frequent breaks along the river. She especially enjoyed playing in the bulrushes, wandering through the mud and looking for frogs, bugs and other creatures that live in the marshlands along the river. She was having fun using her creative imagination—just being a kid: no electronics, no television. There is so much we can learn from the innocence of youth. We never felt deprived as we were enjoying the wonders of nature together and feeling fortunate to be outside on a beautiful day. The Ottawa River is scenic. Its shore, lined by poplar and birch trees, is dotted with marshlands and sandy beaches. We tried to envision the Ottawa River as it was two hundred years ago without today's landmarks. We imagined how early settlers must have navigated this waterway and how vast the land must have seemed to them because we were already finding it extensive. We imagined how beautiful the tall and majestic stands of pine must have been along the river during 1821. They are long gone due to the logging trade of that time.

A valuable source of information was John, Glenn's uncle and one of the members on our 'office support team' back home. Tracking us on his computer and using satellite images, he sent us emails that we could access through our mobile phone. These emails explained, complete with directional arrows on photos from satellite images, possible scenarios for us to manoeuvre around the dams. We could choose which option worked best. John also included 'points of interest' and facts along our route. He provided us with statistical data on our progress. We came to rely on him throughout our entire trip across Canada. Thanks to modern technology and from the comfort of his home, John was living the adventure with us. Before our trip, we had had only four months to prepare, upgrade equipment and acquire items we deemed necessary. We reviewed maps, acquired a Global Positioning System (GPS) and struggled to learn how it worked, but there was not enough time to figure out every paddle stroke and obstacle along the way. So, along with John's assistance, we happily 'winged it' day-by-day. This certainly added to the excitement of the adventure.

We paddled toward Montreal, Quebec, the largest urban centre along the Saint Lawrence River. This city boasts a population of three and a half million. Busy highways crisscrossed the river. Homes and condominiums were built close together along the shoreline—not to mention the industrial sections found in every major city. Like Ottawa, a large city like Montreal presented a challenge to find a secure place where we could set up our tent along the shoreline. In the heart of Montreal, our passage was through the Lachine Canal National Historic Site Lock system. These locks are used for recreational boaters to bypass the Lachine Rapids where the Ottawa River converges with the Saint Lawrence River. Navigation is through five locks that raise and lower the boat traffic to compensate for the elevation change. Individual lock operators controlled the locks.

Each lock operator we met would inform the next lock operator of our impending arrival. The lock staff would have the gate open, ready for us to lock through. On the day we arrived, it seemed we were the only vessel travelling through the locks. However, it did not matter as the lock operators kindly accommodated us anyway. Reaching the last lock at the end of the day and not wishing to proceed into the turbulent waters of the Saint Lawrence River, we decided to look for a spot to pitch our tent. Glenn asked the lock operator where a campground might be. The lock operator informed us there were not any close by as we were in the Old Port, that is, the heart of downtown Montreal. We inquired as to whether we could set up our tent and spend the night on the cement wharf. She informed us that this had never been requested before. She heartily granted permission, although, she said, she would have to charge us a mooring fee for our seventeen-foot six-inch canoe. Asking how much that would be, we learned it was one dollar per foot. We happily agreed and gave her eighteen dollars. Glenn said to keep the change. This location put us in plain view of the many locals and tourists dining on a multitude of downtown patios. We are confident that we were a source of entertainment for onlookers amazed at the amount of gear we pulled out of the canoe to explode on the wharf. We unpacked our bags as we prepared for the night.

This location gave us a rare opportunity to join patrons of a restaurant. Fearing items may go missing, we felt a little apprehensive leaving our canoe and gear unattended while venturing into populated centres. We placed the valuable items in a medium-size dry bag and took it with us. The close proximity of this restaurant, complete with an elevated view, made us feel a bit more comfortable. We were able to enjoy a meal while in full sight of our canoe. Not to our surprise, we were a bit of a novelty as canoe camping was not a usual sight in downtown Montreal. People pulled out their cell phones and researched the name 'Canoe for Change' on the side of the canoe. Offering to buy us drinks and take our photos, they approached us on the patio. Glenn even had a lesson in navigating through the various screens on the GPS. We went to sleep that night to the sounds of the city—sirens blaring, trains squealing, and music and chatter drifting from the restaurant patios.

The next morning, the lock operator and her team prepared coffee for us in the control room and gave us directions on how to travel through the swift currents of the Saint Lawrence River. Small vessels, such as canoes, usually are not permitted to enter the turbulent waters of this ten-kilometre section of the river. However, the lock operator had notified the Canadian Coast Guard and the Montreal Port Authority, and they were well aware that we were coming through. Knowing this, we felt prepared to meet the challenge that lay before us as we ventured off through the fifth and final lock.

Saint Lawrence River
Montreal to Rivière du-Loup, Québec

Dates: August 17 to September 3

Route: Saint Lawrence River

With our gear in place, spray deck zipped tight and energized by strong coffee and adrenaline pumping, we prepared for the turbulent waters that lay ahead. Coming out of the safety of the Lachine Canal we paddled in awe past large freighters and cruise ships that were moored at the Old Port of Montreal harbour. We seemed insignificant against these giants of The Seaway. Once past the vessels, we dug our paddles into the fast, churning current of the Saint Lawrence River. To a lesser degree, these turbulent waters can compare to the Whirlpool Rapids at the base of Niagara Falls. Similar to Niagara Falls there is a large volume of water squeezing through a narrow channel.

A small tour boat filled with about twenty passengers suddenly appeared from nowhere. It was coming towards us at a great speed. The boat came from behind a cement pier at a diagonal angle with its engines revving, preparing to enter the fast-flowing whirlpools of the river. Once the tour guide spotted us, he took the opportunity to engage the passengers in a unison chant through his megaphone. They cheered us on in French and English with gusto. We did not have a chance to wave or acknowledge our thanks, as our hands were steady on the paddles and stroking like crazy. The smiles on our faces could not have been more prominent as the group called out, "Stroke, stroke, stroke! You can do it!"

Eventually, the river became calmer and less turbulent. We marvelled at the fast pace we were travelling with little effort. It was so convenient to have such a swift current and the wind at our backs that we did not notice we were in the middle of the shipping channel. Much to our dismay, a Canadian Coast Guard patrol vessel chased and tracked us down, subsequently notifying us by a loud megaphone voice, "Attention, Little Canoe, you are in the shipping lanes. Please remove yourselves immediately and proceed to shore." Glenn said, "We are in the shipping lanes?" Carol said, "They sure like to use megaphones here in Quebec." Eager to comply, we quickly headed to the north shore of the river to continue paddling. Now that we were sharing the waterways with large commercial freighters, we had to become more cognizant of our surroundings and channel markers. Channel markers are laneways on which lake freighters travel along the river. The markers are hard to spot when sitting in a canoe at water level, but we made a concerted effort to look for them.

Further down the river and several hours later a second Canadian Coast Guard patrol vessel approached. It was on a training exercise and was notified by the first ship to find the 'little' canoe to ensure we were still, in fact, off forbidden waters.

Once we reached Lake Saint Pierre, which is basically a wide area of the Saint Lawrence River, we hauled our wind sail from under the spray deck and harnessed the wind to assist us. Specifically for our paddle across Canada, we purchased this wind sail and used it whenever possible. Glenn remembers ordering the wind sail and having it shipped to his place of work. On the day it arrived, pretending to be in the canoe, he sat in the lunchroom and practised deploying and collapsing the wind sail repeatedly until he felt comfortable that he understood the technique. The light, compact sail has a flexible, circular batten which deploys instantly like a spring-loaded umbrella. Glenn sits at the bow holding up the sail with the bottom tethered to the spray deck in front of him. With a tail or crosswind, the sail moves the canoe effortlessly through the water. Carol at the stern uses her paddle as a rudder to control the direction of the canoe. This method works best in a steady wind of fifteen to twenty kilometres an hour. Too much wind and the boat will become dangerously unstable. Onward we go making good time with our sail.

With the day winding down, we located a beautiful sandy beach. As we went about the tasks of setting up camp, a Great Blue Heron was quietly watching, standing like a statue. As graceful as the heron is, it is hard to believe that these beautiful prehistoric-like birds can emit such a hoarse and harsh squawk when in flight. Under the watchful eye of the heron, Glenn pitched the tent while Carol prepared a pasta dinner. While sitting on the beach eating supper, we enjoyed the backdrop of Lac Saint-Pierre. This area boasts an essential, unique ecosystem for almost three hundred migratory bird species, ninety species of fish and twenty-seven different plant types—what a contrast from the Old Port of Montreal's industrialized area! As we lay in our tent with the fly open, we enjoyed watching freighters pass almost soundlessly, their decks aglow with lights. We fell asleep to the sound of birds and the lapping of water on the shoreline rather than the noisy sounds of the city a few nights previously. We were looking forward to seeing more of this estuary the next day when we would paddle to where the river meets the tides.

Glenn Reflects:

Carol and I continued down the Saint Lawrence River and observed a high frequency of fish jumping out of the water. I said to Carol, "Why would a fish leave its natural habitat to leap out of the water into the atmosphere where fish cannot survive?" No sooner had the words left my mouth when a large fish jumped out of the water directly at me. My reaction was to duck slightly and use my paddle to cross-check the oncoming fish and prevent it from smashing into my chest. Momentum still carried forward: the fish landed on the spray deck between Carol and me where it flopped and then slid back into the water. Carol's reaction from the stern of the canoe was a startling screech. We were shocked, to say the least, but remained composed and in control. I was stunned into silence for the next ten to fifteen minutes as I reflected on 'Did that just really happen'? Naturally, a long string of 'what-ifs' started forming in my mind. 'What if' this approximately thirty-

pound fish had landed in one of our cockpits? How would we remove it? 'What if' it is thrashing about had capsized the canoe? If fish are so eager to get in the boat, next year I am going to bring a fishing rod. Although not an expert on fish species, I believed it was a carp due to its massive girth, colour and barbels near the corners of its mouth. Every adventure has a fish story, and this is mine—believe it, or not.

Past Trois-Rivières, in a narrow meandering section of the river, we found a new challenge—staying clear of lake freighters. These beasts of The Seaway travel at ten knots per hour through the deepest area of the river within the channel markers. Just this year the Canadian government implemented a mandatory, temporary slowdown for vessels in this area over twenty metres or more in length to try to prevent further deaths of whales. Lower speeds give the whales a better chance of surviving a collision. Even travelling at a reduced rate, lake freighters appear to move more quickly than one would anticipate. Our challenge on this day was to stay outside the channel markers. To minimize our paddle strokes and maximize our distance, we would crisscross the river and cut corners by travelling in a straight line through the narrow shipping channel. Well, the inevitable happened. As we were crisscrossing the river 'saving time' we were caught off guard by a lake freighter coming around a blind corner. We quickly paddled out of the shipping channel. Now safely out of the channel and knowing full well that the lake freighter expected to turn, we were in panic mode when it appeared like the freighter was barreling down upon us. As if a swarm of killer bees were chasing us, we paddled frantically towards the shoreline. Once at a safe distance from the lake freighter, we were able to relax. We vowed to be more aware of blind corners in the future. It was indeed a most frightening, yet exhilarating and humbling experience all at the same time.

Marinas were our preferred places to camp. Kind and generous marina operators would give us the use of their facilities by providing mooring for our canoe, a place to pitch our tent and the use of shower, laundry and lounge facilities. They also offered valuable nautical information on river conditions and hazards we would encounter ahead.

Back home, a few weeks before we departed, friends gathered around a social barbeque informed us that the water near Quebec City had significant tides. We glanced at each other as if silently saying "Did you realize that?" Once we got home, Carol had said to Glenn, "Oh my goodness, there are tides in the waters at Quebec City!" With this new information, during the next few days, we frantically researched that tides in the waters at Quebec City could reach up to six metres! This major oversight confirmed to us the fact that we very likely would have more unforeseen surprises along this journey. But is that not what the nature of any adventure is anyway?

This was to be our first experience paddling with tides. Marina operators gave us invaluable advice on how to deal with tidal currents to enable us to travel safely and efficiently. When the water is rising (or coming in) it is called a 'flow tide'. When the water is receding (or going out) it is called an 'ebb tide'. The lull in-between is named a 'slack tide'. At high tide, the water is at the highest. At low tide, the water is at the lowest. There are roughly six hours between these two tides.

We were approaching Quebec City and paddling in the middle of the Saint Lawrence River. Carol was looking at the shoreline and noticed that a particular landmark had not moved for over an hour. Even though we had thought we had forward momentum, we remained stationary. We soon realized we were paddling against the tide. We were in the middle of the river where the current was the strongest. Paddling against the tide is a monumental effort in physical exertion with very little or no gain. When paddling against the tide, we soon learned to stay along the shoreline where the current is less intense. Since we were heading east, we found that travelling in the middle of the river was more to our advantage when the tide is going out.

Paddling toward Quebec City, the high tide for Saint Lawrence River was four and a half metres. Unfortunately for us, when we arrived in Quebec City, the tide was low. This was unlucky for us because when the tide is low, access to the shore is limited or non-existent. We spotted a city park as a possible place to camp. Getting to the shore-line meant we had to beach our canoe in the mud approximately forty metres from land. Slugging over slippery, wet rocks and gunky, deep mud, we had no choice but to carry our gear to higher ground. On this exercise, we exerted a great deal of unnecessary effort. Later that evening, we promised each other that we would plan our future days according to the tides. Therefore, as the days progressed, we would leave two hours before high tide to take advantage of the slack tide. Once the tide started going out, we would 'ride the tide' in the middle of the river and make fantastic progress. A comfortable eight hours later, before the water level became too low again, we would have to look for a place to camp.

The weather was incredibly warm and sunny. Since leaving Ottawa, we had had several days with little wind. The surface of the water was like glass. This made for optimum paddling conditions. During ideal conditions such as this, we felt the need to stay on the water longer to gain distance. Since tides can vary so much, they have a considerable impact on the accessibility of specific areas. It is not uncommon for channels with plenty of water at high tide to become completely dry at low tide. With no marine charts to assist us, we did not realize that we had entered a lengthy stretch of the river we later dubbed the 'mudflat zone'. From what we can gather, because of the estuary, sediment is transported by the tide flow and deposited like dunes on the river bed. Mudflats are exposed when the tide is out.

It was almost five o'clock in the afternoon when we realized we should have looked sooner for a place to camp. The tide was going out and was almost at its lowest level. Soon our paddles were digging into fine silt with layers of mud beneath it. As we advanced, it was a laborious effort until, eventually, we could not move at all. To our dismay, we were aground on a mudflat. There was still approximately five cen-timetres of water on the surface of the mud. We naively thought we could drag the canoe the remainder of the distance to the Montmagny ferry channel, about half a kilometre from our location. Not giving it a second thought, we got out of the boat only to sink up to our calves in the muck. Now firmly stuck, we learned by experience that leaning forward and lifting your heels will free you from sinking further down into the mud. We got back into the canoe, sucked our sandals out of the mud and cleaned up as best we could. So, there we were: high and dry, floundering like two fish out of the water. We had no choice but to stay put in our canoe in the middle of the river and wait until the tide started coming in again.

After an hour of just relaxing and laughing at ourselves at the absurdity of it all, eventually, the tide came in to set us free. Thank goodness it was a calm and sunny day. Now there was enough water to enable us to paddle to the ferry channel. Once at the channel, we took advantage of the deep water and paddled toward the harbour as planned. By the time we neared the harbour, it was around seven-thirty in the evening, and we were ravenously hungry. As we paddled up to the dock, three local fishermen helped us carry our gear up the muddy bank onto a nice, flat, green space where we could pitch our tent. Another fisherman gave us a couple of cold beers and some smoked sturgeon that we could cook for dinner. Delicious! Ah, thank goodness for the kindness of our fellow Canadians! This friendly group encouraged us to secure our canoe to a tree just in case the tide were to come in and wash the canoe away. Sheepishly we wondered if they had seen us stranded in the middle of the river.

As we headed east towards Rivière-du-Loup, the terrain became more diverse. This area of the Saint Lawrence River opened up to a broader channel and was quieter with fewer motorboats. The shoreline was lined with trees of pine and cedar. Sand was now mixed in with rocky shorelines and shoals. Rock islands are numerous and popular for birds to frequent. Moving down the river, we continued to see flocks of Canada geese flying in formation and calling to one another in encouragement. "No," we say, "Do not fly south yet; it is too early!" Migratory birds and snow geese, come to rest in the Saint Lawrence Valley to replenish their energy reserves for the rest of their migration south. The birds find plenty of ingredients for their diets along the coastline, rich and plentiful with bulrushes and aquatic life. It is here that we saw in the water seasonally-installed, long fence-like structures. In Ontario, because of dwindling numbers of eels, commercial fishing is banned. However, a limited fishery remains in Quebec. Eels can be as large as one metre in length and can weigh three kilograms. As we passed by, Glenn hoped that one would not find its way into our canoe. Now as we paddled into mid-afternoon, the wind picked up, as promised, by the marine forecast.

After experimenting over the years with different varieties of water purification devices, we use an ultraviolet light purifier as our preferred method to sterilize water. The process is simple. We dip our empty drinking bottle into the water source, turn on the compact pen-shaped device and immerse it in the water bottle. A water sensor activates the ultraviolet light to sterilize the water, making it safe for drinking. The water sensor removes the DNA of bacteria, viruses, and protozoa, rendering them harmless. It does not remove particulate, but Glenn says that is what the gaps in your teeth do. The downside is that the device requires lithium batteries. On a positive note, the ultraviolet light purifier is quick and efficient. We have even used it while travelling in Asia and South America.

Thirsty, we dipped our drinking bottles into the water to prepare to sterilize it and much to our surprise, we found we had entered into undrinkable salt water. With no drinking water and rain in the forecast, we decided it was time to call it a day. Gratefully, we found a beautiful rocky point to camp on at Rivière-Quelle and pitched our tent behind a stand of cedar trees. They protected us from the impending strong winds. At the unexpected turn of events of not having a fresh supply of water, we devised catch basins to collect rainwater from our tarp. It rained all night and we were able to sterilize ten precious litres of water. We had also left the canoe inverted and collected more rainwater for the days ahead. We usually curse the rain, but in this particular instance, it was a lifesaver. Unfortunately, we had to pack this little device away for the time being as it does not remove salt from water. From this point forward, we had to carry more freshwater that we obtained along the way. We made provisions for this by bringing along a sizeable, collapsible water container.

It was the first day of September, and it had been three weeks since we had left Ottawa. We had been windbound at Rivière-Quelle for two days now and were very close to Rivière-du-Loup. There was a wind warning in effect, making it impossible to launch our canoe in the shoreline surf. As we reflected on the last several weeks, we weighed the chances of being able to make it, as planned, to Sydney, Nova Scotia by the end of October. Even though we were on track, we were impatient and frustrated to be losing precious time while waiting on shore for the winds to subside. While we waited, we reflected on the many challenges we had had to deal with on the Saint Lawrence River: lake freighters, tides, mudflats, rocky shorelines, wind and waves. The timing of the tides had affected our daily routine. Low tides and mudflats had prevented us from accessing the shorelines. Now that we were entering the immense expanse of the Gulf of the Saint Lawrence, wind and waves were our challenges. A sailor had previously warned us of dangerous tidal currents in this section of the river. Tidal currents are caused when massive amounts of water associated with the changing tide are forced around islands and up channels. When this water gets pushed through narrow channels, one can expect strong currents. In some cases, large rapids with towering waves and whirlpools occur in the middle of the river: not the proper place for our canoe.

Finally, after two nights of our being windbound, the wind eventually subsided and we arrived in Rivière-du-Loup. We were thankful to be off the mighty Saint Lawrence River.

Madawaska, Saint John and Canaan Rivers
Rivière-du-Loup, Québec to Shediac, New Brunswick

Dates: September 4 to 21

Route: via Highway 185 to Petit Témis
Bike Trail to Cabano on Lac-Témiscouata;
Madawaska River; Saint John River; Lake Washademoak;
Canaan River; Highway 122; Highway 15

From Rivière-du-Loup on the shores of the Saint Lawrence River, we needed to walk approximately eighty kilometres with our canoe to Lac-Témiscouata, our next water source. Glenn assembled the canoe cart which was stored in the bottom of the canoe. He firmly attached the canoe to the cart using straps. Together, we then proceeded to pack all the gear into the canoe. We learned from the onset that a well-balanced load in the canoe provides stability in the water. This also applies when we are pushing the canoe on the cart, on land. Finding that sweet spot where the load on the cart is balanced was critical. One of us has to push the canoe, and the other has to pull. This method was totally new for us and indeed a learning process: who was going to pull and who was going to push the damn canoe? As with any couple, there are going to be disagreements. Especially for those who spend every waking minute together, day after day, week after week. After heated words, we managed to work it out. We came to a compromise after trying and testing several different scenarios. We rigged up a rope so Carol could be harnessed into the front of the canoe to pull, while Glenn pushed.

With the amount of gear and gadgets we had brought with us, all needed to be organized and stowed in the canoe the same way every

27

time. We accomplished this by packing items in categories. The sleeping gear and clothing are placed into two soft-sided, bright yellow, compression, waterproof bags. We divided the remaining equipment, such as the tent, tarps, small camp chairs, stove, pots and pans, and other camp items between two waterproof bags. One was bright orange, the other blue for easy identification. Food was stored in two bear-proof barrels. One barrel contained breakfast and snacks, the other contained dinner. The gear was always packed and repacked in the same coloured bag every time and loaded and reloaded into the canoe the same way every time. This routine helped us quickly locate gear, thus saving time.

Through downtown Rivière-du-Loup we portaged uphill and continued endlessly uphill through the industrial section. About to embark on county roads heading to our next water source, Lake Temiscouata, we met two cyclists on vacation from Halifax, Nova Scotia. They informed us of a different route called the Petit Témis Bike Trail. They told us the bike trail traces the course of an old railway line linking Rivière-du-Loup to Edmundston, New Brunswick. The trail passes through Cabano at Lac-Témiscouata and follows the shoreline of the Madawaska River. They even gave us their route map. We were so incredibly grateful and felt so fortunate to find this marvellous trail that we could not thank the couple enough. The pair said all we needed to do was call them if we got into a situation—in the water or out—anywhere on our way to Sydney, Nova Scotia. They would come immediately to our rescue. We came to discover these two kind people were just one example of the many generous Canadians we were to meet across this great land of ours.

Much to our delight, the Petit Témis had all the amenities, Quebec style. Explicitly for cyclists and pedestrians, the trail leads across fields and horse pastures into an expanse of calm forests, marshes and lakes. Though it climbs into the hills, the inclines of the trail are gradual, nowhere exceeding a grade of four per cent.

Incredibly happy and grateful for this hidden gem, we stopped at one of the numerous rest areas. While in the shade of a covered picnic table and taking in the view of the pond that lay before us, we enjoyed a lunch of fresh fruit, cheese and bread. Not far away were wild apple trees and Carol gathered fruit that would be a welcome addition to our morning's breakfast of oatmeal and raisins. While Carol was picking fruit, Glenn was chatting with cyclists curious to hear about our adventure and our cause. Not every day had they seen an expedition canoe so far inland. Every time the trail intersected a road, we noticed how hilly the countryside was. This was a constant reminder that we could be walking on the pavement and camping in the ditch had we not found this trail. Eighty kilometres later we launched the canoe at the small town of Cabano on Lac-Témiscouata. This pretty lake is only forty-five kilometres long and five kilometres wide. Later in the day, we took advantage of a covered picnic shelter along the water's edge as a prime place to camp. After experiencing several days of light rain, we were grateful to have shelter from the elements. It is incredible how wet and heavy gear gets even if one does their best to keep everything dry.

Leaving Lac-Témiscouata, while the showers continued, we entered the Madawaska River as it twists and turns snakelike towards Edmundston, New Brunswick. Both sides of the river shoreline through Maliseet First Nations' land is dense with brush and trees. We needed to find a place to rest for the night and this terrain would not accommodate us. Just past Maliseet, we met a local French Canadian who gave us permission to camp at the Madawaska River Airport's access ramp. This grassy ramp allows seaplanes to taxi from the water onto the dry land of the airport. He also told us we had just crossed the border from Quebec into New Brunswick. Excited to have one province, Quebec, behind us, we set our watches ahead by one hour to the Atlantic Time Zone. One province down, seven more to go! Following the French Canadian's advice, we pitched our tent in an open, grassy field right beside the seaplane runway. He did not have any jurisdiction over the airport; however, he knew the airport staff and, if they questioned us, he said to use his name as a reference. We hoped this was not going to be a problem as we were boldly camping behind a large, obvious 'No Trespassing' sign. Glenn conveniently draped his wet rain jacket over the sign as if to hide the words. The rain showers had finally stopped giving us an ideal opportunity to eventually dry out our gear. The night was thankfully uneventful with no planes coming or going.

The following day we headed to Edmundston where we entered the Saint John River. The Saint John River drains a vast ancient mountain area that lies roughly two-thirds in New Brunswick but also includes hills of northern Maine in the United States. Originally the Maliseet named the river 'Wolastoq', meaning 'beautiful river'. And beautiful it is. The terrain changes dramatically every thirty kilometres. We travelled through Grand Falls where the river rages like Niagara Falls and slides through the rich farmland of the Saint John River Valley. We quietly paddled under the 'World's Longest Covered Wooden Bridge' at Hartland. This was our halfway point to Sydney, Nova Scotia, approximately nine hundred kilometres from Ottawa. The river moves park-like through Fredericton's incredible city greenspace. The river twists as it turns its way past First Nations' Oromocto and provides a border between the United States and Canada. As we passed, our vessel was followed and scrutinized by the American Border Patrol.

Under the watchful eye of a bald eagle high up on the cliff, we pitched our tent. We looked up to see the steep sides of the river consisting of clay, shale and rocks—a natural gorge giving us the feeling of a wilderness adventure. Well back from the edge of the cliff and out of sight are houses, communities and potato farms enabling complete, quiet privacy. The river is rocky and shallow. Glenn pitched the tent on a rock shoal of golf ball-sized cobblestone. We wondered if it would be unbearable sleeping on this lumpy surface. Miraculously, our self-inflating air mattresses, thin as they were, gave us the cushioning we needed to provide us with a restful night's sleep.

The next morning we woke up to dense fog, a common occurrence in the Maritime Provinces. The fog had not lifted by the time we had finished our morning's breakfast, nor by the time we packed up our gear and loaded the canoe. We could see no more than twenty metres in front of us. Rather than take the risk of paddling blindly through the set of rapids ahead, we guided our boat through the shallow waters.

We did this, donned in pants to insulate us from the chilly water, by walking beside the canoe. The pants are made of waterproof material and have socks attached, enabling us to stay warm and dry. Once the fog had lifted a few hours later, we jumped into the canoe and traversed the gently rolling rapids. Although the water level was low in the Saint John River due to lack of rain, it was passable, and the conditions were relatively good. At various spots in the middle of the river, we could get out of the canoe and stand. It must have looked odd from the distant shoreline to see two people standing in the water in the middle of the river.

Regardless, 'paddle on, paddlers' we enthusiastically said to each other after several days of a well-deserved break, as we pushed off from Fredericton. It felt good to recharge and rest our tired bodies from the rigours of day-to-day paddling and of setting up and breaking down camp. On either side of the Saint John River near Oromocto, the land consisted of a mixture of farmland, marshland and forests of coniferous and deciduous trees. There were open green spaces, surrounded by the privacy of forests, at the end of abandoned roads and paths. We had inadvertently chosen one of these inviting green spaces to camp for the night. During the night, explosions awakened us. We later figured out that we were close to a military range training area. We confirmed this the next morning as we paddled through the Canadian Forces' Military Base Gagetown. As the explosions became louder and more frequent, Carol said, "I hope they know we are in the water floating through in a canoe." No sooner had the words left her mouth when a military helicopter hovered in the sky above our boat. The occupants were so close we could see them as they leaned out of the helicopter door. We could well imagine the helicopter and its crew using us as a subject for a strategic training exercise. The aircraft soon left, leaving us to paddle in peace through wetlands towards Gagetown Island.

Paddling quietly through the wetlands we saw eagles, osprey, herons, ducks and a multitude of shoreline birds. We noticed banks of mud and realized, once again, that we were affected by tides—this time from the Bay of Fundy. Water levels fluctuate approximately thirty centimetres between high and low tide; not a huge difference, but something of which to be aware. Coming into the village of Gagetown, which is a unique, rather quaint and historic town, we said goodbye to the Saint John River and turned the corner into Lake Washademoak. This lake eventually leads to the mouth of the Canaan River. Much to our dismay, we found the water in the Canaan River low and gradually getting more shallow. We hoped we could make it further into the Upper Canaan River but had our doubts. We paddled approximately one more kilometre, and our fears were confirmed. For as far as we could see, there was nothing but exposed rocks and boulders, making the river impassable. Intending to get more information on the condition of the river, we reluctantly decided to turn around and head back towards a Recreational Vehicle Park we had passed the day before. The park owner put us in contact with his friend at Fisheries and Oceans Canada to get information on the status of the Canaan and Petitcodiac Rivers. These two rivers were our intended passage to Shediac on the Northumberland Strait. The conservation officer said the Canaan River was too shallow due to severe drought conditions experienced this summer and fall. He also strongly recommended that we not enter the Petitcodiac River with a canoe. He said that once into the river, we would be caught in a tidal bore and would not be able to climb up the steep muddy banks to get out. A tidal bore is a phenomenon in which the leading edge of the incoming tide forms a standing wave that travels upstream against the current. At high tide, the extraordinary volume of water from the Bay of Fundy floods into the river. The river banks narrow and the compressing waters rise in a spectacular surge and visible standing wave sometimes one metre in height. Generating wakes in its path over three metres high, this churning water races upstream at speeds close to fifteen kilometres per hour. With this new information there was no way we would paddle;

so overland we had to travel again. Eventually, we made it to Shediac on the Northumberland Strait.

Northern Northumberland Strait
Shediac, New Brunswick to Pictou, Nova Scotia

Dates: September 22 to October 1

Route: Northumberland Strait

It was now past the middle of September, and the end of October was less than six weeks away. The most challenging part of our journey lay before us. We were now on the salty waters of the Northumberland Strait. This was a new experience for both of us as we had no previous experience with ocean paddling.

Shediac, New Brunswick is known as the world's lobster capital for its lobster fishing, processing plants, live-lobster tanks and the legendary Lobster Festival. We launched our canoe near the world's largest lobster monument. We gazed at the enormity of it as people gathered around taking 'selfies' of the ninety-tonne structure.

Oh, the vastness of the Northumberland Strait left us apprehensive of what lay ahead. The enormity of the Atlantic Ocean waters would be the real test of our new canoe. We had upgraded from a smaller, heavier canoe to a more extended, lighter model, especially for this trip. It was a 'Prospector', specifically designed to be comfortable. It features increased buoyancy to float over large waves rather than to go through them. Just over seventeen kilograms in weight, the Prospector is composed of advanced composite Kevlar and carbon materials. From the experience of having owned a 'Prospector' configuration in the past, we have come to know it is a stable boat, a proven Canadian classic design and can hold its weight. We had a rugged spray deck installed to provide added protection from the elements. With the cover, we had hoped to gain aerodynamics and, at the same time, added protection from any waves washing over our canoe. As an added benefit, perhaps not necessary, the cover kept the gear out of sight.

On our first day on the Northumberland Strait, our anxiety was alleviated somewhat by being blessed with calm waters. We spotted a gem of a spot to camp for the night and quickly beached our canoe on the shore. This beautiful cove had fine white sand and was surrounded by large smooth rocks that rose to a high embankment the size of a double-decker bus. We noticed there were several homes a reasonable distance away. Beaches, the locals say, are free for everyone to use and belong to no one; above the waterline or embankment is where property ownership starts. We pitched our tent in a spot where we did not think the ocean tides would affect us. One can tell where the high tide water line is by the seaweed that has washed up onshore. Also, the sand appears dry and not saturated with water. With camp set up and wanting to unwind from our first day on the ocean, we sat down on our collapsible chairs while savouring a cup of coffee infused with Irish Cream. Taking in the view, Glenn said: "This is better than a big-screen television." While relaxing on the beach, we met Sandra and Daniel who were beachcombing for coloured glass. Sandra uses glass to make jewellery and artwork. They nicknamed this cove 'Paradise Beach' and said it was a special place for them. Sandra gave us a smooth piece of olive-green beach glass and said it was a rare find. Much to Glenn's delight, Daniel gave him a couple of cold beers and said, "You should be okay where the tent is, I do not think the tide is going to come up any higher than that." So off to bed we went, anticipating a good night's sleep. Sometime after nine o'clock at night, the wind and the waves picked up unexpectedly. We peeked outside the tent as we felt uncomfortably close to the water. Listening to the crashing of the waves, we contemplated Daniel's words 'should be okay'. Carol said, "Daniel only said 'should' not 'is okay.'" After much debate, at eleven o'clock, we donned our headlights and set to work to relocate our tent and all the gear to higher ground. It took about thirty minutes to carry, piece by piece, in the dark, up the steep rocky embankment to a grassy knoll. With camp safely set up again, having peace of mind in knowing that we would not get washed away, we slept much more soundly. The next morning, we noticed that where our tent originally sat, the beach was dry. We had not needed to relocate after all. Lesson learned: think twice before we set up so close to an ocean shoreline.

Joining the province of Prince Edward Island to the mainland of New Brunswick is the Confederation Bridge. We were in awe as we paddled towards it and appreciated the enormous engineering and construction feat necessary for its creation. We discussed how the Confederation Bridge is symbolic as it helps unite provinces. From the shore, we noticed that the bridge is not straight but curved. The engineers designed the bridge this way to ensure that car drivers remain attentive, thus reducing the potential for accidents. At thirteen kilometres long, the Confederation Bridge is the world's longest bridge over ice-covered water, reaches forty metres above the water and rests on sixty-two piers. Seals surrounded us while we paddled between the piers. We presumed it was an excellent fishing spot for them. Breaking the surface of the water and looking about before they ducked once again beneath the water, they were a constant source of entertainment for us.

For many hours before our trip, we had plotted and mapped our course across Canada. Software on our personal computer allowed us to view and organize maps. We plotted various waypoints and set routes and tracks. Once we felt we had the data complete, we uploaded our information onto our newly-acquired GPS device. We are both new to this type of technology and not fully versed in using it to full capacity. In the past, we generally used a compass and map. We preferred these for our canoe trips. However, due to the length of the cross-Canada canoe trip, we decided to obtain a GPS.

There came the point in the journey where we had to deviate from our planned route. However, we did not know how to use the GPS as a compass—a simple task. As mentioned, we have always been comfortable using a topographical map and a manual compass. Still, for a journey of this magnitude, it would have been impossible for us to carry the encyclopedia of paper maps required. So Glenn purchased a road map to accompany our GPS because, he felt, just as in life, he always has to look at the big picture. Our initially planned route would take us along the shoreline of Baie Verte, south of Cape Tormentine. However, with the weather being so incredibly calm, we discussed whether we should take a risk and cut across this vast expanse of water instead. Carol checked both the marine radio and the weather application (app) on her mobile phone.

The weather looked favourable with no winds in the forecast. To leap from the point of Cape Tormentine over to Amherst Beach Provincial Park in Nova Scotia was a span of fifteen kilometres. This decision would save us two or three days of paddling around the bay. After long minutes of debating as to whether or not we should take the chance and try to figure out how to use our GPS as a directional arrow, Glenn pulled out his trusty old compass and said, "I knew this would come in handy—just in case." With our compass heading focused on the park, we had our minds made up and knew full well that this was a considerable risk if the winds picked up. Nevertheless, we paddled hard and were eager to get across. We paddled past colourful lobster buoys set by fishers of the sea to locate their traps. Lobster fishers would race in their trawlers from one buoy to another inspecting their lines that reached far below the surface of the water to the traps. We even paddled out further still. Knowing that we were an uncomfortable seven and a half kilometres from either shore, we stroked more quickly at the halfway point. Focused on our heading, we arrived within a few degrees of our destination in three hours. We were fortunate the winds remained calm and counted our blessings on a successful crossing. Once we reached shore, we hugged each other in happiness and relief that we had saved ourselves so many kilometres of paddling. And, just like that, we crossed over from New Brunswick to Nova Scotia. Another milestone.

It just so happened the Park Warden of Amherst Beach Provincial Park drove up in his golf cart to inspect the beach. He was a bit surprised to see us. We informed him where we came from, and he just shook his head in disbelief. The Park Warden said that to walk with our canoe to the main campground facilities would be a significant undertaking. Instead, he said, we could stay under the apple trees by the hiking trail just behind the beach. It was a nice quiet spot. Carol took photos of the stark contrast between the beach before the tide came in and then, six hours later, when the tide went out. Amazingly, three-foot tides make such a difference in the topography of the landscape. Where there had been water previously now had transformed into rippled flats of red sand. An area where seagulls had bobbed in the water was replaced by a pod of seals basking in the sun and relishing the calm of this remote stretch of shoreline.

The locals had informed us that it is unusual for the Maritime Provinces to experience such hot, dry weather at the end of September. Twelve hours after our calm crossing, it appeared things were going to change. The next morning, we awoke to find that there was a wind warning in effect. From Amherst Beach Provincial Park, we had no choice but to walk along Tyndall Road toward Northport.

We had brought along extra water bladders in anticipation of not being able to use our ultraviolet light purifier with salty ocean water. We obtained our drinking water mostly from day-use parks or from knocking on doors. People were always willing to help us out. These contacts also allowed us to chat about our cause. Transporting water in the canoe while on the water is not a problem but it increases the weight we have to push while walking.

We usually swam daily in freshwater to bathe ourselves, keeping our hygiene presence tolerable. However, on the Northumberland Strait, it was difficult to submerge our heads while swimming in saltwater—especially for Carol. She found that her long curly hair became dry, brittle and stiff and formed masses of tangles like dreadlocks. To overcome this, we still took our swims; however, once back at camp, one of us would wash with soap while the other poured a pot of freshwater over our head to rinse off. We were also finding the saltwater very demanding on the equipment. The zippers on our spray deck became more stiff. Also, the sun seemed to cake and dry the salt deeply into the teeth of the zipper. Our clothing was also getting coated, crusty and stained with salt.

Paddling along the east coast, we soon learned that if we were too close to shore, waves tended to be closer together and crested from the shallowness of the water. Further out, yet still a safe distance from shore, paddling was more manageable as the waves are further apart.

Much to our delight and entertainment, as we travelled down the Northumberland coast we found an abundance of curious seals once again playing peek-a-boo with us. Occasionally they would swim very closely to the canoe. They usually snuck up behind us so we could not see them. Then, as if on cue, they would make a large splash as they caught us off guard. Every time it made us jump! We are used to beavers doing this manoeuvre on our backcountry canoe camping treks. We could not help but wonder if the seals and beavers had the same purpose in mind: to startle humans on purpose.

During these past few nights, after settling into our sleeping bags, we noticed a loud droning noise. Glenn related it possibly to stock car races in the distance while Carol connected it to a herd of cows bellowing. It was not until a few days later that we realized this sound was from hundreds of grey seals gathering to breed on remotely exposed sandbars in bays along the coast. Nightly, for many hours, we heard the bellowing over a great distance.

Days later, with our confidence level high, we took the challenge and crossed another large expanse of open water. This time it was across Tatamagouche Bay. After this crossing, the winds picked up. Not wanting to get too close to the rocky shore in the cresting waves, we anxiously scoured the coastline for a safe place to land. Fortunately for us, a sandy point came into view. There was a field of wildflowers behind a hill. It was an accommodating spot for us to set up our camp for the night and to protect us from the ever-increasing wind. That evening our tent began to shake, straining the flexible poles from the force of the wind. Thankfully the tent was pegged down securely, and miraculously the symmetrical frame maintained its integrity. We had chosen a four-person backpacking tent specifically for this trip. With the extra room of a vestibule at each end, the tent easily accommodates the two of us and two of our gear bags. This three-season tent packs up into a smaller bundle than most two-person tents and weighs less than three kilograms. The tent has a meshed ceiling covered by a separate waterproof tent fly. The tent fly can serve as a temporary shelter from a bad storm and can be erected alone. With so many tents to choose from on the market, we had chosen well.

We continued paddling down the coastline towards Caribou Point. The winds were not proving to be in our favour. Rounding Caribou Point, we would have been more exposed to the elements of the open Atlantic Ocean. Plus, we would have been competing with ferry boats leaving the Northumberland Ferry Dock to Prince Edward Island. So, with two reasons not to travel beyond the point, we decided to walk inland with our canoe on its cart. This walk would be a mere fifteen kilometres along Highway Six to calmer waters. Looking at the positive side of the situation, we realized walking would save us considerable time and effort rather than if we were to paddle around the point. It was getting late in the season, and we still had a considerable distance to reach our destination of Sydney.

We were walking along the roadway to our next destination, the town of Pictou. At the edge of a property, we came upon a bench. Inviting passers-by to stop and rest while at the same time enjoying the beautiful view it offered, the bench was much like what you would see in a park. Taking advantage of the opportunity to rest our feet and have a snack, we had the pleasure of meeting Bob and his neighbour Kevin. Bob was a retired fisherman from the area. He kindly gave us his phone number in case we ran into trouble. He said he would pick us up if we ran into trouble anytime or any place. Bob was well-versed in the coastline and gave us tips on what we could expect on the paddle ahead. He warned us specifically about Cape George that could be, depending on the weather, 'either sugar or shit'. "Cape George, Nova Scotia, which is past the protection of Prince Edward Island, leaves one exposed and open to the full force of the Atlantic Ocean," he said. Bob also provided us with a mobile phone number of the Canadian Coast

Guard. In the meantime, Kevin had run to his home and brought back two bottles of homemade blueberry wine. Carol thought she had died and gone to heaven. Off to Pictou we marched, looking forward to a glass of wine, or rather a 'cup of wine', from our mess kit.

Southern Northumberland Strait
Pictou to Saint Peters, Nova Scotia

Dates: October 2 to 9

Route: Northumberland Strait

After several weeks of paddling on the ocean, we booked ourselves into a bed and breakfast in Pictou. The thought of a hot shower after weeks of sponge baths tickled our imagination. Just the idea of washing the dirt, salt and stickiness off our bodies made Carol giddy. Pictou is also the first community we had seen along the coast since leaving Shediac, New Brunswick. It was a timely opportunity to replenish our breakfast and lunch supplies. Pictou is one of those gems of a town you stumble upon and where you could stay forever. The town is renowned as the 'birthplace of New Scotland', as it was here, that the first wave of Scottish immigrants landed in 1773. Pictou offers beautiful period homes on tree-lined streets and bright, multi-coloured buildings along the harbour waterfront. Buildings are painted in vibrant, bold colours of yellow, blue, red and orange so they could easily be noticed in former days by fishermen coming home from the sea.

While relaxing in Pictou, Glenn took the opportunity to get his hair cut by a local barber. As Glenn was sitting in the chair, the apron neatly fastened around his neck, he was startled to see the barber put on two pairs of glasses, one on top of the other. Not wanting to offend the barber and resisting the fight-or-flight response, Glenn did not move a muscle for fear of losing some flesh. As we were walking down the street, we met Bob again. He had been driving all over town searching for us to return our water bottle which Carol had left while sitting on his bench. Bob was excited to have located us and exclaimed, "Glenn, you got your haircut since I last saw you." Chuckling, we thanked Bob profusely for returning our water bottle.

While in Pictou, we took the opportunity to launder the salt out of our clothes, replenish our diminished food supplies and enjoy some local cooking. Preparing to leave Pictou, we promised each other that someday we would return to visit this beautiful town. We retrieved the canoe stored behind the bed and breakfast, placed it back on the canoe cart, loaded our gear and headed to Pictou Harbour. Once at the harbour it was now the routine to disassemble the canoe cart, place the canoe in the water and reload the gear. We left Pictou, heading towards Cape George, against a strong easterly wind. During the past few days we had enjoyed the sight of pods of seals following us. They were ever curious to see what was encroaching upon their territory. They would playfully sneak up behind the canoe, break the surface of the water and slowly turn their heads left and right to survey their surroundings. At one point a grey seal with long white whiskers and a drawn face appeared beside the canoe. The seal had a striking resemblance to our elderly neighbour George—lacking only round spectacles perched on its nose. It is comical how your imagination wanders when you have long periods of time to reflect.

We marvelled at the beauty of the coastline with its heavily-treed forests and dramatic cliffs in the forefront made of sandstone and shale. Now and then we saw fishing harbours but they were few and far between. We had to proceed carefully along this rugged shoreline of jagged rocks and cliffs and always be on the lookout for a place to land at a moment's notice. Fierce winds creating huge ocean swells too big for a little canoe can pick up quickly during this time of year. Surely enough, the inevitable happened when, days later, the weather radio warned of gale-force winds by noon. 'Gale-Force Wind Warnings' on coastal, inland and offshore are issued when wind speeds of thirty-four to forty-seven knots occur or are expected. 'Strong Wind Warnings' are issued when wind speeds of twenty to thirty-three knots arise or are expected. What it does not include are the gusts which can add to the fury of the situation. With the impending Gale-Force Wind warning in effect, we left very early on this particular morning to make at least some headway. We had our sights set on Lismore Harbour about twenty kilometres down the coast.

Surely enough, by eleven o'clock, the winds increased. In situations like this, when paddling on the ocean becomes dangerous, Carol had learned to embrace the swells. Steering the boat in 'sync' with the waves, not against them; to become one with the waves and not to fight them. Much like a dancer does with their dance partner. We were struggling with swells as big as cars. Waves were starting to crest, washing over our canoe. "What have we gotten ourselves into?" we thought to ourselves as we struggled through the cresting waves to manoeuvre our canoe which was loaded down with over one hundred kilograms of gear. With Glenn in the bow and Carol at the stern, Glenn said, "Let me know when we are not having fun and we will turn around and go home." We knew we would have to get off the water soon as, with each passing hour, the wind was continuing to build.

As we continued along looking for a safe haven where we could land, to our amazement, a sandpiper dropped out of the sky from nowhere to land on the spray deck directly in front of Carol. Seconds later we noticed a hawk circling overhead and surmised it must have been chasing after its prey, the little bird. We were comforted to know that we could offer some refuge for the sandpiper who seemed to be tired and in obvious distress. Carol could see the little creature as it scrambled more closely, breathing heavily, seeking her protection. She felt a motherly bond with the sandpiper and felt the need to protect it from any harm as it huddled more closely in the embrace of her defence. Seeking safety from the increasingly steady rising of the wind and waves, we continued along for at least another hour. We spotted what we thought was a breakwater entrance to Lismore Harbour. With the seas being as turbulent as they were, it was difficult to visualize that there could be an opening among the towering cliffs of this section of the ocean coastline. As fate would have it, we were second-guessing ourselves only to realize there was indeed a narrow opening. Inadvertently we had lost the opportunity to enter safely into the harbour. We had already passed the point where the direction of the waves would take us safely into the refuge. We discussed whether or not we should risk turning around in the heavy sea and head back out to make another attempt or continue down the coastline. We made the joint decision to attempt to go into the harbour. This attempt meant turning the canoe around between the swells and approaching the entrance again. It was going to be a "Hang on. We are going for it," moment. We knew it would take a great deal of skill and timing to control the canoe to avoid being hit broadside by a cresting wave and risk capsizing. With adrenaline pumping, we completed the manoeuver. But we lost our little companion. A wave splashed over our deck and forced the sandpiper to fly away and seek its own safety in the trees.

As we silently said goodbye to our new-found friend, our second attempt into Lismore Harbour was a scary ride. The situation had our pulses racing. We were both concerned about capsizing. We managed to position the canoe such that the tailwind and direction of the waves lined up into the port entrance. From the stern of the canoe, Carol 'threaded the needle' as the boat rode on the top of a cresting wave. Mere metres from the breakwall, we lunged forward through the harbour entrance. Having narrowly missed the rocky cliff entrance on either side, we breathed a sigh of relief to be in the protection of this well-hidden cove. It was instant calm. One would never know we had left a raging ocean behind us. Passing a colourful array of lobster fishing boats, we paddled to the end of the harbour. We tied up at an empty spot at the dock. Grateful to have averted what could have resulted in a disastrous ending, Glenn climbed out of the canoe and kissed the ground.

Thankful to be off the water before Mother Nature released all her fury, we decided to walk the remaining eight kilometres to Arisaig Provincial Park. A couple in a pickup truck stopped to ask if we needed assistance just as we were out of the harbour and now walking along a quiet country road with our canoe. They said they had been watching us paddling by from their ocean-view home. The man was a fisherman by trade and said he would not have been out in those conditions in his fishing boat, let alone in a canoe. He said they had seen us rise in the swells and then disappear, hidden at the bottom of the waves. They had continued to watch us for several minutes until we disappeared. Concerned about our safety, they had come to look for us to ensure we had made it to shore. They asked us if we needed assistance or a ride to our next destination. We politely declined. They were relieved that we were okay and headed back home. It is truly a wonderful feeling to know that strangers are genuinely concerned about the well-being of their fellow humankind.

Arisaig Provincial Park is a day park. Like many parks along the east coast, it does not offer overnight camping. However, for two people pushing a canoe down the highway or paddling through on the waterway, a day park complete with covered picnic tables, privies and hiking trails to explore is a great place to camp for the night. There is always a flat grassy spot to pitch a tent. Park staff made an exception for us to overnight camp due to autumn being a quiet tourist season. Wondrous Nova Scotia where people are so much more relaxed and laid back. Makes you want to sell everything you own and move there.

Much to our dismay, the next morning, the winds were blowing hard again. We woke up refreshed but still in mild shock from the previous day's events. There was no way we would go out on the water to attempt to paddle around fearsome Cape George, especially after yesterday's near capsize. So with the canoe cart, we started walking the thirty kilometres to cut across the point to Antigonish Bay. This was our longest portage in one full day. We did walk eighty kilometres from Rivière-du-Loup to Cabano, Quebec, but that was over four days. As we were walking along the road, a vehicle slowed, the window rolled down and in a comical tone, the driver stated, "I think that canoe works better in the water!" We both waved and forced a smile, each of us secretly thinking of our own colourful retorts.

With the weather unseasonably warm, including little rain and hours of sun, we were thankful for the ideal camping conditions. Places like Fredericton and Edmunston were recording record high temperatures. High winds are typical on the eastern seaboard but this year of our voyage, according to 'mariners of the sea', had been more blustery than usual. This year had one of the most hyperactive Atlantic hurricane seasons as well as the costliest tropical cyclone season on record. Nevertheless, we pressed forward, regardless of the conditions. Of course, paddling is our preferred method to complete the journey, but we had prepared to walk—as long as we did so under our power. The extra unexpected exertion might be causing us to be quick-tempered and frustrated. Still, we were trying to take everything in a positive stride.

Carol Reflects:

Glenn and I were starting to feel the effects of the significant amount of physical energy expended to undertake a journey of this magnitude. It was taking a physical and mental toll on our bodies. It had been fifty-five days since we had left Ottawa and we had both lost weight. Our faces were becoming drawn, our thoughts foggy, and our bones were beginning to show. Glenn had made a belt from twine to hold my pants up; he also had to tighten his belt a couple of notches. I was preparing the morning 'breakfast of champions'—oatmeal with raisins. As Glenn was coming out of the tent, he said, "I must be hallucinating, I can smell bacon and eggs cooking." We had planned to add fresh vegetables and fruit to our prepared meals along the way; however, on our route, there was hardly any opportunity to do so. I came to realize that the meals did not have enough calories to sustain us versus the calories we were burning. Our diet lacked the necessary fat, fibre and whole grains that we needed to stay healthy. Not having enough calories equals weight loss. To say the least, after weeks of rotating the same five meals, we became overly sick of the lack of variety. Plus, I found that my digestive system had been in turmoil for the past few weeks. Since I was unable to eat appropriately, I felt weak and unable to do my share. I felt guilty because Glenn was taking on more of the physical workload of portaging. Despite the challenge, staying positive is extremely important. If not for yourself, at least remain positive for your partner. There is nothing worse than having your paddle pal complain, sulk or be moody. Thank goodness we are compatible and were able to have fun despite the rigours of the journey.

Staying motivated while pushing our canoe up and down the hills of the Maritimes could also be a challenge. However, on the positive side, it gave us a chance to use different muscles, to see the beautiful rolling hills of the coast and the occasion to meet people. It also gave us a perfect opportunity to chat about our cause—to talk about food insecurity in Canada and how those we met were affected by food insecurity. Meeting people and receiving their encouragement and vocal support for our cause and our journey inspired us! Several people come to mind. One is a young woman named Sharleen who noticed us walking by and came running out to chat. Once she heard our story, she turned to call out to her mother and father to meet us. Before long, her neighbour came out as well–practically a small street party. After that lovely chat, we continued on our way only to have both Sharleen and her mother come back with a nice loaf of freshly baked bread, home-grown tomatoes and cucumbers. Sharleen had looked up our website and saw that we were promoting food, in particular fresh, healthy vegetables and fruit. We were speechless at this offer of generosity. It gave us extra bounce in our step as we continued down the road.

Later that day, we encountered construction along the highway. The recently-paved surface bore no shoulder and left a dangerous drop-off of sloping soft gravel, thirty centimetres deep. We had no choice but to walk directly on the windy road with countless blind corners. We made ourselves as visible as possible by draping our brightly-coloured rain jackets over our shoulders. A short time later we were stopped by a gentleman, Greg, who asked us, "What the heck are you doing walking on the road in a construction zone? It is far too dangerous!" He insisted that we load our gear and canoe into the back of his truck so he could drive us safely through the construction. Along the way, he invited us to his home and insisted we join his family for Thanksgiving dinner. Carol and I turned to each other, with an expression of surprise and said, "It is Thanksgiving?" We were so focused on our journey that it had not even occurred to us: it was a holiday weekend. The mere thought of a turkey dinner with all the fixings and a home-baked pumpkin pie made our mouths water. How could we turn down this east coast hospitality? Greg and his wife Triena invited us to use their spare room, have a hot shower and use their laundry facilities. In the evening, we went to the home of Greg's mother, Linda and her husband Larry, who had prepared a sizeable, delicious turkey dinner! They had no idea how much we craved this timely nourishment. We tried to be polite and not to shovel the food into our mouths. We were very grateful for the heartfelt and warm reception that their family so graciously extended to us. They made us feel like we were part of the family. It renews one's faith in humanity when kindness is spread from one person to another in completely innocent trust. We were so very, genuinely thankful.

Bras d'Or Lake
Saint Peters to Sydney, Nova Scotia

Dates: October 10 to 18

Route: Bras d'Or Lake; Atlantic Ocean

Not far from Greg's house, at the small town of Saint Peters, we entered into beautiful Cape Breton Island. Connecting the Northumberland Strait to Bras d'Or Lake is the Saint Peters Canal. Bras d'Or Lake is a large body of both fresh and salt water at the core of Cape Breton Island. The lake measures roughly one hundred kilometres in length and fifty kilometres in width. To get to Bras d'Or Lake, we had to pass through the canal. The canal is a single lock featuring an unusual double-gate system. There is tidal activity at both ends of the canal. At one end the Atlantic Ocean has a tidal variation which is entirely different from the Bras d'Or Lake at the other end. For this reason, both entrances have double-lock gates. The tide was in when we went through the lock. This made the transition approximately a thirty-centimetre drop from one body of water to the other. If the tide was out, it could have been a one hundred forty-centimetre drop.

Cape Breton Island is one of Canada's most enchanting places where the land meets the sea. The island is at the eastern extremity of the Gulf of Saint Lawrence. Even though the land surrounding Bras d'Or Lake is not as steep as the northern part of the island, large hills encircle it. As we paddled along the shoreline, we were awestruck by the spectacular beauty of the trees that were at the height of fall colours. We enjoyed the lake's sheltered coves, secluded beaches and sandbars making Bras d'Or Lake ideal for canoeing.

On hot and humid days, we learned to expect high winds. True to our expectations, as soon as we left the protection of Saint Peters Canal, we worked against the wind all day until we had no choice but to land at an island not far from the mainland. The island looked towards the mainland where we could see a large hill with a narrow road leading down to the shoreline. There was no way we could cross the channel from the island to the mainland as the wind was too strong. The island was a bit of an oddity to us with its one hundred and thirty colourful shacks scattered around the southern tip. We also found the island littered with all types of debris, everything from children's toys to broken lawn chairs. It was like people had left in the middle of the night with no time to pack up their belongings. One building in particular that stood out was a church. With a small cemetery behind it, the church was made of wood and painted white. By erecting the tent on the side of the building facing the mainland, we used the church to shield us from the wind. We walked around the island to do some exploring. The days were getting shorter with daylight turning into cold darkness around seven o'clock.

Extremely tired from paddling in a headwind all day, we finished dinner and sought the welcome protection of our tent. Winds continued to blow, and gusts shook our tent and rattled the shutters on the shacks. The vision of the leaning iron cross atop the steeple crashing down upon the tent remained with us all night. As the night progressed and until the wee hours of the morning, headlights from vehicles were shining on our tent. The automobiles were high on the mainland. Once vehicles crested over the hill, the lights would shine on the tent. They illuminated it as well as the church behind us. We had no idea why there was so much traffic.

The next morning the lake was filled with whitecaps as far as we could see. Because it was too windy to venture back out onto the lake, we proceeded to pack our canoe and paddle the short distance across the channel to the mainland. Once at the shore, we found a sign with the name of the island: 'Chapel Island Potlotek First Nations' Reserve'. The inscription said that Chapel Island was one of the meeting places for one of seven Mi'kmaq First Nations' districts in the Atlantic Provinces. The island is a national historic site of Canada. Burial sites, dance circles and other features attest to its prolonged use. Erosion of the shoreline revealed bones over four hundred years old starting to be exposed to the elements. Temporary measures had been put into place to prevent further destruction of this site. The grand council of the Mi'kmaq still meets twice a year on Chapel Island, a sacred place for the community. The traditions of the meetings have continued from the eighteenth century and now thousands come to celebrate the feast of Saint Anne to renew their family faith and reinforce family ties. In the mid-eighteen hundreds, a Catholic Church was erected on the island. Even after four hundred years of European influence, its people speak and hold on to Mi'kmaq beliefs, values and traditions. Once we had finished reading the sign, we felt terrible that we had trespassed, intruding on sacred land without permission.

After we had hauled our canoe and gear up the hill to the road, we met a First Nations' citizen. We told him that we had camped at Chapel Island and he said he knew. The entire village knew. Well, that would explain the procession of vehicles! We sought forgiveness if we had offended. He said it was perfectly fine and he was happy that we had found shelter. As we continued pulling our canoe up the hill, we met others who waved and came over to chat with us. They, too, all knew we had been at Chapel Island. Then we met the Chief who had also heard about us and had come looking for us. In speaking with the Chief, we realized that Potlotek First Nations was a close-knit community and were not surprised at how fast news had travelled. No one was offended that we had used the historic land of Chapel Island for temporary shelter.

The Chief proceeded to tell us that the thriving community is having significant problems with their water supply. The Potlotek First Nations have had to endure decades of dealing with brown, smelly water, unfit for drinking due to high levels of iron and manganese. We were surprised that fresh drinking water was not available to this community, with no resolve, for so many years. How could this be in a country like Canada? We can manage to put an oil pipeline through First Nations' lands. Still, we cannot provide fresh water to a First Nations' community in Nova Scotia? The Chief told us his father was still alive. In June 1967, fifteen Mi'kmaq paddlers, in two war canoes, had embarked on a historic sixteen hundred kilometre, thirty-one-day journey from Potlotek (Chapel Island) to Montreal, Quebec to mark Canada's Centennial celebrations. His father was one of them. The Chief invited us to a local hall to view a commemorative display covered with photos and recounting this event in history.

It was the middle of October and there was evidence the seasons were changing as the days were getting colder, more seasonable. The hours of darkness now exceeded the daylight hours. It had been over nine weeks since we had left Ottawa, and Sydney was just days away. The night before we were to arrive in Sydney was the coldest yet with a low temperature of minus five degrees Celsius. During the night our water bottles had frozen shut, forming a thin layer of ice on top. To be comfortable at breakfast, Glenn took down the tent and left the fly up to give us some protection. Our lightweight camp chairs fit perfectly under the shelter and, with warm winter hats and gloves on, we enjoyed our morning coffee with breakfast.

On day sixty-seven, we arrived in Sydney Harbour after a day of paddling on relatively calm waters. We were elated! As we reached the dock, we handed our camera to a utility worker and asked him to take our photograph as we paddled in. We unloaded the canoe for the last time this year amid the hustle and bustle of downtown Sydney. Around the harbour, cruise ships were docked, and tourists were venturing out in amphibian sightseeing buses. We chatted with locals and tourists alike and even had a teenager take a picture of us doing a jump-shot. We were both so worn out we could barely get off the ground, making our jump-shot a pathetic photo. Glenn's excuse was it was his first time doing a jump-shot, but we both knew it was from lack of energy.

We had prearranged accommodation in downtown Sydney. Here we would wait for Carol's brother Paul and his wife Denise, members of our support team, to make the two-day drive from Kingston,

Ontario, to transport us and our gear home. From the harbour, we proceeded to push our canoe to the downtown core. We were looking for a restaurant to have a celebratory lunch. We found there was absolutely no room in front of an establishment on the main street to park a canoe. However, we did spot a Co-operators Insurance office on the corner. We walked over and asked if we could leave our canoe in their parking lot behind the office. Tom, the owner, and his staff could not believe we had just completed our paddle from Ottawa. Once we had convinced him of our feat, he offered us the use of the lower level of his home while we waited for our ride home. We thanked Tom but informed him we had already committed to pre-booked accommodation. Tom insisted we have the use of his wife's car so we could have the freedom to tour Cape Breton Island. Tom walked us to the parking lot and handed us the keys to his wife's BMW Z3 convertible. We were humbled and astonished by his generosity and trust.

Now with a sporty car parked out front, our accommodation was nicely centralized in Sydney. We spent the rest of the week cleaning up our equipment, doing laundry and sightseeing. We also appreciated being off the water. It allowed us to rest and to reflect on the first part of our journey, now complete.

It was also a chance to reflect on the whole past year. The year 2017 was momentous for both of us. Our son Paul was married and one of our daughters gave birth to our third grandchild Juniper. In January, we decided upon a trek in the desert. Craving an offbeat adventure, we elected to hike Cottonwood-Marble Canyon in the heart of Death Valley, California. Once we had pitched our tent in the sand at Stovepipe Wells, Death Valley basecamp, we set about preparing for our six-day hike. Shortly after supper, in the distance, we saw a low cloud approaching. It was merely an observation at that point and we naively never gave it a second thought. However, when it reached us, to our surprise, it was sand. We were in the midst of a sand storm. Never having experienced this phenomenon, we had no idea what lay in store for us. With the rental sports utility vehicle parked close by, we decided to get a good night's sleep, in preparation for our hike the following day and quickly crawled into the tent. This would soon pass, we believed. Much to our alarm, the wind became more substantial and, when peering out, we saw that we were engulfed in swirling clouds of dust and sand. The strong, dry wind blowing across the desert carried clouds of sand and dust dense enough to obscure the setting sun and reduce visibility to zero. Worse yet, the sand entered through the fine mesh of the tent's screen. We burrowed ourselves deeply into the silk liners of our sleeping bags, covered our mouth and nose with balaclavas and hunkered down, waiting for the storm to pass. There was no getting away from the heaps of fine sand collecting within the tent. Suddenly, the pegs securing the corners of the tent gave way and the tent collapsed. Struggling, with the tent flapping wildly around us, we managed to roll up our air mattresses and sleeping bags, folded up the tent and madly dashed to the car. We turned on the headlights to see how bad the storm was and spotted tumbleweeds rolling across the desert—along with one of our sleeping mats. Once we had retrieved it, we slept in the hatchback while the force of the wind rocked the car.

The following morning, with sand in every crevice of our bodies, we set off on our trek. Besides our gear, we each had to carry nine litres of water with only one chance to replenish from a desert spring. The hike took us through three different canyons, we wound through dramatic narrow passages, travelled over steep ridges and trekked across classic desert landscapes. The scenery was stunning.

After our desert hike, we washed our favourite hiking clothes in the outdoor camp sink, cleaned the grit and dirt out of the car, had a shower, then went to Las Vegas and got married. It was a fun and crazy day, totally unplanned. At the world-famous 'Little White Wedding Chapel', we did a drive-through ceremony through the 'Tunnel of Love' which is much like a drive-through at a fast-food restaurant but a little fancier. Not to be delayed further, and after a couple of days of luxurious accommodations, we put our hiking clothes back on. Addicted to the magic of the desert, we continued forward to hike for a week where the Mojave and the Colorado Deserts came together at Joshua Tree National Park. But that is another story.

As enjoyable and stunning as our hikes were, we had gained a new appreciation for the bounty of freshwater—large amounts of it—another good reason to paddle across Canada.

Year Two: Vancouver, British Columbia to Fort Frances, Ontario

May 1, 2018 to September 29, 2018

Pacific Ocean and Fraser River
Vancouver to Hope, British Columbia

Dates: May 1 to May 11

Route: Fraser River

It is 2018 and we were off again! On this leg of the journey we were on our way to Vancouver. We placed our canoe and gear on the Canadian Via Rail Train. We intended to paddle back home to Kingston, Ontario. Once we had arrived at Union Station in the heart of downtown Toronto, we frustrated many taxi drivers as we occupied the taxi lanes with our vehicle. With the help of Carol's second cousin, and Paul and Denise, we unloaded our gear as quickly as possible. We hoped our intrusion would not be an inconvenience to travellers as we occupied this space. Union Station directs and controls the most extensive and busiest rail corridor in Canada. Massive limestone roman Tuscan columns support the main entrance façade, inviting travellers into the great hall where they gather daily by the hundreds of thousands to find their station platforms. Our first portage was through the station. Repeatedly saying, "Excuse us, coming through, and please watch your head," Carol led the way and guided Glenn who only had a limited view of the shiny, pristine marble floor. No one gave us a second look! After all, this is Canada. We blended right in. Why would someone not come through Union Station with a canoe over their head?

We were looking forward to four days of relaxation on the train, complete with meals and accommodation. Our accommodation was an upper and a lower berth, not 'First Class', mind you, but certainly more superior than sleeping in a tent. As luck would have it, we had the whole back half of the train car to ourselves. We enjoyed spectacular views from the observation car, live entertainment and meeting friendly people from all over the world. After our first full day and night of rail travel, we awoke the next day to find out that we were still in our home province of Ontario!

Glenn Reflects:

As I looked out of the window, I thought to myself, what I had gotten myself into? Can I physically paddle all this way back home? After all, I am not a spring chicken anymore. The total railroad distance is 3,360 kilometres—it would be nearly twice that many kilometres to paddle back on the waterways. The lakes are still covered in ice now, but the ice will have melted by the time we reach Ontario. Were we giving ourselves enough time to complete the journey back home this year? Wondering, are we able to walk the four hundred kilometres while pushing a canoe over the Rockies? I swallowed my fear and thought we would go as far as we can; after all, it is not a race but a retirement cruise. I was determined to sit back and enjoy the journey while it lasted, whether in the train or on the water paddling.

We felt keen and very excited to embark on the adventure that lay before us. Last year our paddle from Ottawa to Sydney, Nova Scotia taught us valuable lessons. We had made a few minor changes to our gear, replaced worn equipment but, most importantly, made changes to the food we brought with us this year.

Carol reflects:

Because we had both experienced rapid weight loss and a lack of energy towards the end of our trip last year, I knew I had to make a change in our canoe trekking diet. Over the winter months, I conducted extensive research to become informed on what types of food to consume to maintain health and vitality during the rigours of our journey. It was important to me that we continue with our well-balanced diet full of vitamins, minerals, and substances that we were used to at home. I also had an underlying medical issue. Over the winter, my family doctor referred me to a gastrointestinal specialist who diagnosed me with 'Celiac Disease', a gluten insensitivity. The discomfort I felt towards the end of our trip last year would explain my extreme weight loss and lethargy. Last year I ate more glutenous foods than I usually do which produced an overdose of gluten. I never dreamed it would have such a detrimental impact. Therefore, I came up with a new plan. I eliminated gluten and added more of the nutrients we were missing like unsaturated fats such as avocados, fish, nuts, seeds and cold-pressed vegetable oils like olive oil. Included were ideal proteins for maintaining muscle like beans and legumes, eggs and dairy, sun-dried tomatoes, seeds and soy. Also added were energizing complex carbohydrates like brown rice, oats and quinoa, sweet potatoes, fruit and vegetables. I set to work over the winter and spent weeks dehydrating the food in our dehydrator. I dehydrated fresh food like bananas, apples, carrots, beets, sweet potatoes, cabbage and mushrooms, to name a few. There was one instance when Glenn donned a snorkel mask to keep him from tearing up while chopping ten pounds of onions with uniform consistency. I even went as far as dehydrating items like chickpeas, homemade tomato sauce, olives and yoghurt. These were the ingredients I felt were necessary to prepare delicious meals sprinkled with a light vinaigrette dressing, spices or seasonings. Gone were last year's rotating five prepared meals without the essential nutrients. We had learned good food could make a difference between an enjoyable trip and a disappointing one. Now our meals at times were creative, always flavourful and, when possible, were made with food foraged from the forest or meadows. Wild leeks, smooth rock tripe (lichen), blueberries and even edible wild rose petals were added for presentation. Considering the intensity of the journey, I estimated that we would each consume about four thousand calories per day or twice as much as we usually do. This preparation took a lot of planning. It involved a lot of food to get across Canada. It was impossible to carry all that food with us so we prepackaged supplies ready to be sent out to us by friends to 'yet-to-be-determined' drop-off spots. I knew it would be a challenge to keep our energy stores up and I still expected we would lose some weight, however not to the extent that we had last year.

After days on the train, we finally arrived in the early morning in Vancouver, British Columbia where yet another portage awaited us through Pacific Central Station. Once out of the train station, we marched through the streets of Vancouver to our first water source—Creekside Park. Once on the water, we paddled out of the harbour and passed high-rise condominiums, office towers and markets until we reached the Strait of Georgia and the shores of the Pacific Ocean. Basking in temperatures of twenty degrees Celsius, combined with stunning views of the Vancouver Island Mountain Ranges, we were happy to be on the water once again.

Our destination was the mouth of the Fraser River between Vancouver Island and the most south-western point of British Columbia. We had intended to paddle up the Fraser River to the town of Hope.

Before we left on this leg of our journey, friends and well-wishers had been concerned with how we were going to find a multitude of waterways from Vancouver back home to Kingston. "How are you going to get over the Rocky Mountains?" they would ask. Other modern-day voyageurs travelling across Canada by canoe start at Rocky Mountain House in Alberta which is on the east side of the Rockies in Alberta. Then they travel down the Bow River towards the South Saskatchewan River in Alberta. The voyageurs start in Alberta because the watershed flows east from the Continental Divide of the Rockies. From the Continental Divide, in the opposite direction, the water flows west towards the Pacific Ocean. Right from the start, our goal was to travel from coast to coast on our own steam. Weighing our options and, after careful consideration, we came up with a game plan. Once we arrived in the Town of Hope, we were going to travel by mountain bike over the Cascade Mountains through the Okanagan Valley to the next water source—the West Arm of the Kootenay River in The City of Nelson. There we would retrieve our canoe and continue paddling. Before we left home, we had prearranged the purchase of mountain bikes and panniers through a local bike shop in Hope. The plan was to place our canoe and some of our equipment into storage. With our gear minimized we would bicycle on the Kettle Valley Rail Trail and the Columbia and Western Rail Trail to the Kootenay River. These trails also encompass the 'Great Trail'. Once a comprehensive railroad system, the decommissioned tracks are now home to an extensive recreational trail providing almost six hundred fifty kilometres of connected pathways throughout the region. To cross off another adventure on our bucket list, we would do a little sightseeing on the way at the same time.

Our immediate challenge was paddling against the current. We found it quite manageable at the mouth of the Fraser River; however, as the river narrowed, the flow became faster. Further upriver we had to compete with tugboats towing huge barges carrying wood chips, sawdust and fuel to pulp and paper mills. We had not expected the Fraser River to be so industrialized. Not only was the traffic on the river congested but the shoreline consisted of forestry mills, rusty shipyards, steel industries and train yards used explicitly for the lumber industry. When possible, we paddled between log jams, and the shoreline, along the edge of the river. Here the current was the weakest and was out of the way of heavy river traffic. Regardless of how hard we paddled and how determined we were to make progress, we became alarmed that at the end of three days we had only travelled forty-five kilometres. We had expended so much energy and had gained so little distance! Once at Surrey, we realized there was no way we could paddle against the current for one hundred twenty kilometres to the town of Hope.

At Barnston Island Ferry Dock we hauled our canoe and gear out of the water. While at the ferry dock we watched with curious fascination as a small car ferry travelled back and forth across the river. The ferry was just a simple pontoon moved by a little tugboat. It did not travel in a straight line across the river but was angled sideways, then backwards, then forward with the strong current. It was a proficient manoeuvre, and we admired the skill of the tugboat captain. Just watching the tugboat struggle against the current confirmed to us the impossible task of trying to paddle any further. So, not to be defeated, we arranged a taxi van to pick us up, along with our canoe and gear, to transport us to Hope. Once in Hope, we would complete this section of our route by paddling with the flow of the current back to Surrey.

Nestled among mighty peaks at the confluence of three rivers, Hope has some of the best scenery Canada has to offer. The mountain peaks were gloriously encrusted in ice and snow with the rushing rivers below. We pitched our tent at a small campground at the Fraser River's edge. We had a spectacular view of the snow-capped mountains as a backdrop. The first thing we noticed was how incredibly fast the river was flowing. The following day not only would we have to launch our canoe in the fast current but we would also have to contend with mountain debris from spring snowmelt. We somehow would have to paddle while manoeuvring the canoe around ice chunks, logs, branches and whole trees floating in the water. So much debris, in fact, sometimes created log jams. As much as we enjoyed the beautiful scenery, we went to bed that night apprehensive and laid awake worrying about launching our canoe in the fast current of the river. Tired after a fitful night of sleep, we carefully and somewhat fearfully, launched the canoe back into the Fraser River and headed downstream towards the City of Surrey. Indeed, the river was fast but not as furious as we had imagined. Along with all the free-floating natural debris of the river, we rode the current with little effort. We relished not having to paddle against the current. We enjoyed the gift of spectacular mountain scenery—what a stark contrast to the industrial section of the lower Fraser River of Greater Vancouver. After a day and a half, we arrived back in Surrey, thus completing this portion in our canoe on our own steam. Our taxi picked us up to take us, once again, to Hope where, using a completely different set of muscles, we would transition from canoeing to bicycling.

Rail Trails

Hope to Nelson, British Columbia

Dates: May 12 to May 30

Route: Crowsnest Highway; Kettle Valley Rail Trail;
Columbia and Western Rail Trail

With our canoe in storage, we set out to cycle in near thirty-degree Celsius conditions. We were excited by the prospect of cycling on the Kettle Valley Rail Trail and experiencing some of the fifteen tunnels and eighteen restored wooden trestle bridges. We quickly found out that a series of old train tunnels, the Othello Tunnels, was closed for construction. Disappointed to miss this group of historic tunnels, we were resolved to the fact that we would have to cycle along the highway and intersect the trail later.

Cycling along Crowsnest Highway just outside of Hope, we found small, neatly-arranged piles of trash that had been collected and placed on the shoulder of the road. The highway crews must be out executing spring cleanup we thought to ourselves. When we stopped for lunch near a pleasant-looking stream to get out of the sun, Carol spotted a middle-aged man and offered a friendly hello. Together, the three of us converged and sat down to have lunch. We found out that his name was Eric, and it was Eric who had been collecting the roadside trash. Volunteering his services in this particular way is Eric's way of giving back to Mother Earth for all that she has provided for him. Eric is a homeless man and comes from a broken background. He had lost his son recently at the age of eighteen from a Fentanyl drug overdose. We sensed Eric to be a humble man who speaks well of others and cares passionately for the earth. We shared our lunch with him and tears of gratitude welled up in his eyes. Such a simple gesture means so much to a person in a difficult situation. Eric is a Christian man and believes God will provide for him. We shared our stories and found out all he wanted out of life was to provide for his grandson, be self-sufficient, have a piece of land on which to grow food and build a meagre home. We discussed and shared our beliefs about the importance of a food-secure Canada and talked about issues surrounding food insecurity.

Food security is not always an easy subject to explain. Canada does not have a lack of food. Food insecurity in Canada does not look like starvation in Africa. We, in Canada, are less apt to see distended bellies but more likely to see obesity because of the low food quality that is available to people. Type II diabetes, depression, asthma and fibro-myalgia are correlated with food insecurity. Individuals should not be judged by these outcomes—it is a systemic problem. We are 'canoeing for a change'; a change to promote the importance for all Canadians to have accessibility, affordability and the right to fresh food.

Our mountain bikes, burdened with gear and food, felt bulky and cumbersome on the paved road. Both of us found our leg muscles pro-testing against the weight and challenge of cycling in the mountains. Throughout the Cascade Mountains, we had to walk up hills that were too steep for us to cycle. We both hoped to reach the rail trail where the incline grade is only a little over two per cent.

Back on the Kettle Valley Rail Trail we camped peacefully along the decommissioned tracks, drank from the freshwater mountain streams and enjoyed breathtaking scenery. We cycled through farmlands, ranches, lush forests and on the edges of splendid, steep-sided canyons. It is not hard to picture the feats of engineering that went into blasting through these canyons and imagining the romantic days long ago when trains with puffs of steam trailing behind snaked through the canyons. From the top of wooden trestle bridges, one can marvel at the rock cuts intertwined with numerous rivers below as the water descends through the hard rock, sharp curves and steep grades of three mountain ranges. The trails also led us through quaint little towns, cities and communi-ties such as Grand Forks, Christina Lake and Castlegar.

It is at the City of Grand Forks we saw the effect of significant spring flooding that had devastated residents along the rivers. The flooding was a result of considerable snow accumulation in the mountains. As the temperature soared so did water levels everywhere in the Kootenay area. As a result of this snowmelt the Granby and the Kettle Rivers, which converge at Grand Forks, flooded the entire countryside. From our vantage point, high on top of the mountainside trail, we were astonished to see, down in the valley, farmland, buildings and roads submerged. The rail trail went through Grand Forks and along the river where we could see residents had piled debris at the side of the road. Outside their homes lay their possessions that had been waterlogged and damaged. Our hearts went out to the inhabitants along the river for the devastation they had to endure from the floods. We wondered what else lay ahead as we cycled past. We soon found out that, in some areas, washouts caused by the swollen river removed portions of the trail as if a tsunami had come through and had swept the trail violently away. At other times we had to take off our shoes and socks to walk through cold water where washouts on the trail made it too deep to cycle through. At higher elevations, snow covered the trail and made cycling cumbersome. Before we left Hope, we had had visions of a firm, well-packed surface of finely-crushed gravel winding through the mountains. This was most often not the case. Many areas encompassed long sections of sand, coarse stone, potholes, landslides and ruts caused by all-terrain vehicles.

On the Saturday of the Victoria Day long weekend in May, we camped near First Nations land. We found a patch of grass beside the trail, complete with a bench and a picnic table that served us very well for the evening. The following morning, as we were packing up our gear, we were surprised to see three magnificent wild horses on the trail watching us. Standing from a safe distance, they curiously eyed our tent then glanced beyond us and slowly turned their heads to look further down the trail. That gave us a clear indication they wanted to travel past our tent but were reluctant to do so. As we slowly and methodically packed up our gear, Glenn removed tent stakes and calmly talked to the horses. Nodding their heads and snorting made us think they understood what he was saying. The instant the tent poles came down and the tent had collapsed, all three horses, as if acting on a prearranged plan, dashed past us only to stop, turn their heads and look at us as if to say 'we made it'! Not only did we have the rare opportunity of seeing these beautiful creatures, but we continued to be amazed at the incredible beauty of the area.

We cycled through the Okanagan Valley, one of the hottest and sunniest spots in Canada, where we needed to be on the lookout for rattlesnakes sunning themselves on the path or nestled under a rock. We marvelled at the lush and beautiful countryside as we passed manicured orchards of cherry, peach and apple trees—not to mention elegant vineyards. With rows of vines carpeting the slopes below the trail we passed wineries that are world-renown. Winding along the trail as we headed towards the City of Penticton, we saw vineyards stretching from the lake to the sky: a truly visually-stunning vista. Of course, we could not go through the Okanagan Valley without stopping to explore some of the vineyards to have a glass of wine and drink espressos at a local coffee shop. In Penticton, we were treated by Eric and Rebecca, children of dear friends back home, to a superb home-cooked supper. On a trip such as this, people have no idea how important it was to have loved ones provide care, support and encouragement as we continued on this long and adventurous journey.

The landscape changed into rolling hills, sagebrush and pine as we reached a milestone along the Kettle Valley Rail Trail—Trout Creek Bridge. The steel-truss bridge stretches one hundred and eighty-eight metres long. Standing seventy-three metres high, the bridge is also the tallest on the Kettle Valley Rail Trail. This engineering marvel was, in its time, North America's highest steel-truss bridge. When builders laid down the final span, it came within a pinky finger's width of perfection, a testament to the chief engineer's skill.

Not only did we marvel at the magnificent bridges along the trail but also its multitude of tunnels cut into the sides of mountains. One, in particular, is Bulldog Tunnel on the Columbia and Western Rail

Trail. The tunnel is nearly one kilometre long and has a curve blocking the light at the end until one is halfway through. Not wanting to go further this day, we set up our camp on the west side of the tunnel. At the entrance of the tunnel there was a steady stream of cold fresh mountain water cascading down from the stone archway. Hot and sweaty from days of cycling, we frolicked like two naked children as we ran underneath the cascading water. With stunning views of the forested valley and river far below this piece of paradise was our most scenic camping spot on the rail trail system. The following morning pushing our bicycles, we donned our headlamps and started walking into the dark tunnel. Our headlamps cast an eerie light on the tunnel as we proceeded cautiously, carefully manoeuvring our way around puddles and bits of slag along the trail. The ceiling of the tunnel still revealed blackened stone from the soot of steam trains that had passed through many times. Once we got past the bend in the tunnel, we could see the end.

Two weeks and four hundred kilometres later, we arrived in The City of Nelson and made arrangements to retrieve our canoe. We were eager to get back onto the water and were looking forward to putting our canoe into the West Arm of Kootenay Lake.

We felt gratified and rewarded that we had had the unique opportunity to see a majority of the historic Kettle Valley and Columbia and Western Rail Trails. Another item knocked off our bucket list!

In Nelson, we had an opportunity to have press coverage with the Nelson Star, a local newspaper. A few weeks earlier, Glenn had done a radio interview with the Canadian Broadcasting Corporation (CBC) Radio giving our cause good promotion. Still nervous at giving talks, we are resolved that it is necessary to raise funds and build awareness around food security. We hope the more we do, the easier it will become. With social media and added press coverage, we wanted to direct people to our website so they would be inspired to donate to our cause. David and Debbie from DFC International Computing have built and maintained our website, a most generous gift. Website online donating circumvented our need to carry a tin can full of cash

71

and cheques. We made the website as attractive as possible by uploading photos, camp recipes, equipment lists and newspaper articles. Our daughter Rachel designed a blog for our website and updated it weekly. Carol would send her pictures and all the trip information through texts or emails, and Rachel would make it come alive with her writing skills. She spent an enormous amount of time and energy to make us look good.

With the constant sending of information came the increased drain on the batteries to power our devices. To assist with keeping our gear charged, we had brought with us a compact solar panel. This small folding solar panel laid flat on our spray deck and enabled us to recharge our batteries.

Rocky Mountains
Nelson, British Columbia to Pincher Creek, Alberta

Dates: June 1 to June 19

Route: Kootenay Lake to Creston;
Creston to Fort Steele (Via Crowsnest Highway);
Fort Steele to Kootenay River;
Kootenay River into Lake Koocanusa;
Lake Koocanusa to Elko (via Crowsnest Highway);
Elko to Pincher Creek, Alberta (via Crowsnest Highway);
Pincher Creek to Oldman Dam on Oldman River
(via County Road #785)

Now through the Cascade Mountains and the Okanagan Valley, we entered Kootenay Lake which is surrounded by the Selkirk mountain range on the west with Purcell mountain range on the east. The lake is a widening of the Kootenay River which rises in the Rocky Mountains and flows south into the United States before looping sharply north back into Canada. We would not be entering the United States but would pick up the river later. Although the time on this lake would be short, we were happy to get our paddles wet. On this long, deep and fjord-like lake we had one of the most spectacular campsites of our entire journey complete with cold spring water, a sandy beach and a view of the mountains—with no civilization in sight! As with all good things, the short time spent on this lake came to an end. We headed for the east shore towards The Town of Creston. It is here that we began our long and arduous journey from British Columbia to Alberta with only a few chances to be on the water again.

Soon after we got off Kootenay Lake, Uncle John, back home, sent us a news story of a kayaker who had been attacked and mauled by a grizzly bear in the area we had just left. Oh, that was close. We wondered what had provoked the attack. There is always a reason as, generally, grizzly bears usually avoid humans. One of the most common circumstances associated with a grizzly bear attack is the 'sudden encounter,' especially with a female grizzly bear and her young. Perhaps the kayaker had surprised the bear while walking behind the campsite. We are vigilant and on the lookout for bears, especially when on the trail or choosing a camp to settle in for the night. When walking through the forest, we tend to make noise, sing or talk to each other to avoid a sudden encounter. We practice no-trace camping and always make sure our camp area is clean and tidy. We had received a lot of teasing before we embarked on our trip. Friends, neighbours and relatives all had suggestions as to how to remain safe from grizzly bears. Tips ranged from packing a gun to bear deterrent spray to flares; all of which we determined would be unnecessary weight. Our niece's husband gave us a highly-animated impromptu grizzly bear lesson. His experience from tree-planting training in northern Canada embellishes the different characteristics of grizzly bears versus black bears. One of his noteworthy facts was 'once a grizzly bear attacks, and he starts to lick you—it means he is going to eat you.' Reviewing all this good advice and information, we opted to take a lightweight air horn. At the end of the trip, we were not even sure it worked as we never had a chance to use it. Whenever we saw bears, the horn was packed away with our tent in the gear bag. It is like seeing the perfect photo opportunity, and your camera is out of reach.

Walking on the Crowsnest Highway through the Kootenay Mountains we were in and out of a few waterways, such as Moyie Lake, Kootenay River and Lake Koocanusa. We were awestruck by stunning views as we paddled with the current in the valleys of these waterways between snow-capped mountain ranges. It is here that we had the rare opportunity to spot a cougar from our canoe. It was a fleeting moment as the sleek animal, perhaps down at the shoreline to get a drink of water, sensed our presence. We rounded the corner in the river, and the cougar quickly made himself scarce. Lake Koocanusa is a reservoir in British Columbia and Montana and is formed by the damming of the Kootenay River by Libby Dam. Lake Koocanusa's name is derived from the first three letters of Kootenay, Canada and USA. It would be the last time we dipped our paddles into the waters of British Columbia.

We spent the next several weeks walking over the Rocky Mountain Continental Divide. The distance we travelled every day varied depending on how steep the grades of the mountains were. It also depended on how many hills there were each day. Our march over the Rocky Mountains along the Crowsnest Highway into Alberta seemed daunting. Still, we stayed positive and took it as an opportunity to enjoy the mountains at a slower pace rather than to whiz by in a car. The Crowsnest Highway, for its entire length, follows the alignment of the Canadian Pacific Railway and passes through communities such as The Town of Elko, The City of Fernie and The Municipality of Sparwood in British Columbia. We would walk to the summit of Crowsnest Pass Highway. Once we crossed over the height of the Continental Divide, we would walk downhill into Alberta, past the towns of Frank and Pincher Creek and the Hamlet of Lundbreck Falls.

We had a few comical moments walking along the highway as well as a few scary moments. More than once, we had to walk our canoe and cart through 'Watercraft Inspection Stations'. The inspection stations are set up by the Alberta government to check recreational boats for invasive water species such as plants, zebra and quagga mussels which could potentially travel from one water source into another. It is mandatory for everyone transporting watercraft to stop. The staff were amused when we came walking up with our canoe on the cart.

We had great fun when we had the opportunity to enter a highway weigh station, where to determine their weight, trucks and commercial vehicles go over the scales. Fortunately, it was during lunch hour, so we did not hold up the proceedings too long. Here was our opportunity to weigh our canoe and gear. We were surprised by the fact that we were pushing and pulling our fully-loaded canoe at one hundred ten kilograms; much more weight than we had thought. It was no wonder this march along the hot pavement caused painful blisters on our feet. Each morning we tended to the affected areas by applying moleskins and blister band-aids. Carol was nursing six blisters on one foot alone and eventually lost two toenails. So painful were the blisters we would alternate between walking in hiking boots and walking in socks and sandals.

To relieve the constant strain on our backs, we devised a pulling system. We outfitted Carol with the food barrel harness minus the barrel. We tied each end of one long rope to the left and right side on the front frame of the canoe cart. Once connected firmly in place, the rope crisscrossed under the bow of the canoe and then fastened securely to Carol's harness. Carol felt like a horse pulling a wagon.

Each evening, we set up our camp as far as possible from the road. Provincial rest and picnic stops were great places, but when not available, it was not uncommon to find us in a ditch or a field. We often envied cyclists, who in the course of a day, were quickly travelling from one designated location to another. As we were pulling and pushing our canoe up a rather long hill, we came across two couples cycling across Canada. They were covered in coal dust, black, with soot from head to toe. In the District of Elkford, coal dust is on the shoulder of the highway. The coal dust is deposited by transports carrying coal from a nearby open mine pit. Unfortunately for these couples, the wet, showery weather made the situation very dirty. Fortunately for them, they had a motorhome travelling with them. The cyclists were always provided with meals, snacks and drinking water by the driver of this support vehicle accompanying the foursome. Glenn said, "Now that is the way to go! A dry bed and meals prepared for you every night, not to mention a glass of wine and a hot shower at the end of each day." In our conversation, we referred to them quite often in the upcoming days. Whether it is walking, paddling or cycling, we are all having the adventure of a lifetime. We all gauge our experiences according to the passions we desire. The important thing is that we are doing it. Along our trek across Canada, whether we were portaging or on the water, people would ask, "Is it not dangerous?" It is not the dangers of the adventure we worry about but the danger of growing old someday and not having experienced life. Reaching the final stage of our life's journey and saying 'I wish I had done this or that,' is not where we want to be.

A further example of living one's life dream is a group of three cyclists we met. A young mother, over a span of two years, was giving both of her two sons the education of a lifetime. They had started in Quebec City and cycled to Florida, had crossed the southern United States to California and cycled up the west coast into Canada. They were now heading east, back home to Quebec. The children were the young ages of seven and nine—no doubt a memorable experience for the whole family. Here was a unique opportunity to gain an education firsthand and not only connect with nature but have the chance to learn about North America's rich culture and history.

For its entire length Crowsnest Highway is a two-lane highway with a paved shoulder. The shoulder is a little over one metre wide. After the young family had passed, it was an hour or so that we came to a sharp bend on the meandering highway. A guardrail was in place to prevent vehicles from careening off the edge of the mountain into the narrow valley of the Elk River, far below. We were overtaken by a pickup truck—complete with flashing lights and caution flags—that was travelling at a high rate of speed. The truck served as a support vehicle, warning the oncoming traffic that a 'wide load' was approaching. We were trapped in the worst possible spot—between guardrails and the edge of the highway. There was no time to run to an area that would be a safe distance off the highway. As we backed against the guard rail with our canoe, we could see the transport was towing a trailer with an industrial excavator. The excavator was considered a 'wide load' as its huge wheels extended well past the sides of the trailer. The transport truck driver was racing to keep pace with his support vehicle. We could see that he was shocked to see us. At the last moment, he quickly veered his truck into the oncoming lane. As a result, the truck's load swayed dangerously from side to side. We stood back in horror, leaning over the guardrail as far as we could, while the wheels of the excavator flew by, narrowly missing us. Fortunately, there was no traffic in the oncoming lane or we would all have been in one heck of a wreck. The team was travelling too fast and it appeared the support vehicle had not warned the transport driver that we were there. We immediately thought of the mother and her two boys who were ahead of us and out of sight and prayed that they would be okay.

On one occasion, in Elk Valley before Fernie, we found camping in the ditch to be not all that bad. We conversed with a road worker performing highway maintenance, and he warned us of an upcoming tunnel. He cautioned us to be careful about it. He also recommended a potential camping spot just off the highway and several kilometres past the tunnel. As we continued down the highway, the tunnel loomed ominously ahead. Judging that there was a pause in the traffic, we sprinted through the dark narrow passage safely to the other side. Luckily, we did not encounter a 'wide load' transport or cars racing to get to the other side. Once through the tunnel, we were on the lookout for an abandoned road beside the current highway. The area recommended to us by the road worker turned out to be a pleasant spot despite being close to the highway. The road, once a paved surface, was cracked and covered with tangled shrubs, weeds and loose gravel. We pitched our tent on the abandoned road directly under the barrier of the highway that reached approximately ten metres above us. Being protected by highway guardrails above lessened our fear of being squashed by a falling vehicle. Ignoring the noise of the traffic high above us, we had a beautiful vista of the valley below, complete with elk grazing along the river.

We had thought we were relatively unseen in this spot tucked down below the highway but were taken aback when a slow-moving pickup truck approached us. As it turned out, the occupants were a lovely couple scouting the area for landscape rocks to install in their garden. After a short chat to find out what on earth we were doing with a canoe camped beside the highway, they took us in their truck to go on a quick sightseeing tour of Fernie. Fernie is a charming little city which lies amidst peaks of the Rocky Mountains in the narrow valley of the Elk River. These peaks form part of the Continental Divide. Before returning us to our tent, Lorraine and Ralph took us out for dinner and gave us once again the opportunity to speak about our favourite charity. Much to our delight, the opportunity turned into a generous donation being made towards our cause.

We had started this portion of our trip at sea level and had been climbing gradually uphill ever since. In short order, we would be crossing over the Continental Divide, also known as the Great Divide. Our trek along the Crowsnest Highway had given us many opportunities to chat with locals and view the countryside. We determined our rate of progress to be approximately four to five kilometres an hour. Not only did we have the chance to speak with many Canadians along the way, but walking gave us a glimpse of what the Rocky Mountains have to offer in the form of rugged and spectacular views. We had seen very little rain since leaving Vancouver. It started to sprinkle, and we looked up in surprise and thought, what is that coming from the sky? Not that we, personally, wanted rain; however, much-needed rain was in order as the province was on alert for another year of potential forest fires. As we progressed towards the top of Crowsnest Pass, Alberta, our canoe and equipment became burdensome as it had become water-laden from an afternoon shower. We took the first opportunity we could find to erect our tarp over a picnic table at a roadside rest area and changed into dry clothes to stay warm. As we climbed in altitude towards Crowsnest Pass, the temperature was falling. Now we could expect more rain as mountains attract clouds which generally produce more condensation than in the valley below.

Up, up, up we went pulling and pushing the canoe until we reached Crowsnest Pass summit at one thousand, three hundred, fifty-eight metres. Crowsnest Pass summit is also the point where one crosses over from British Columbia into Alberta. Just like that, we were at the top of the Rocky Mountain Range. It had taken us thirty-nine days to cross the province of British Columbia. From this point, we would be walking down from the Rocky Mountains into the flatlands of the prairies. Here, at the Continental Divide, is where the rivers born in the Rockies flow eastward to Hudson Bay or westward to the Pacific Ocean. We were delighted to reach this momentous milestone in our trip where the rivers now flow in the direction we were travelling.

We had known the province of British Columbia was going to be a daunting challenge for us, but we were happy that we had been able to encompass it in our paddle from coast to coast thus making a real paddle/portage across Canada. We had traversed five separate Rocky Mountain Ranges: the Cascade, Selkirk, Purcell, Monashee Mountains and finally the Continental Divide of the Rocky Mountains into Alberta. Our route was approximately twelve hundred kilometres to the Oldman River in Alberta from Vancouver in British Columbia. In summary, we estimate that we paddled four hundred kilometres, bicycled four hundred kilometres, and walked four hundred kilometres; forty-six days in total.

Just before the Information Centre in Alberta, we saw a small herd of mountain goats leaping from rock to rock along the steep walls of the pass. We marvelled at their agility and had fun watching them bounding through the mountains without a care in the world. The scenery of the mountains is even more spectacular in Alberta. With many lakes and rivers along the road, each step we took would bring us closer to our next water source, the Oldman River.

We viewed many beautiful mountain vistas with every step we took across Alberta's Rockies. On the other hand, we must have been quite a sight for motorists—especially commuters who travel back and forth to work. When commuters would first spot us, they would slow down and curiously look at what we were doing. In response, we would wave our arm to greet them, as we do with all those who pass us. As they returned, later in the day, we would be still walking, and now they slowed down and would wave back to us. The next morning, they would find us again walking along the highway, but now they would wave and smile. As days rolled on, we were recognized by the same commuters as they honked their car horns and waved enthusiastically. The number of these occurrences was ever-increasing the more we walked; repeating the ritual like a predetermined pattern. Soon commuters would stop, and, like old friends, ask where we were going. The next day they would bring us coffee or treats like cookies or fresh fruit.

Now in Alberta, we felt a sense of accomplishment with another province behind us. The landscape was changing drastically. With the higher peaks now at our backs, the sky ahead of us appeared endless as we entered 'big sky country'. The transition between the elevated heights to foothills of Alberta seemed welcoming. Down, down, down we went into the valley and the wide-open plains of Alberta. We now saw fields of crops, large farms with more silos and grain bins than one could count; after all, this is prime fertile land. As we walked along the highway, we saw a string of turbines harnessing the wind and the entire length of a train. From engine to caboose, the train stretched across the great plain; something that one does not see in many parts of Canada!

One never saw two weary individuals, such as we were, dancing for joy to see the Rocky Mountains behind us. Not to mention how we looked forward to seizing the opportunity to cool off our blistered feet in the waters of the Oldman River. For many weeks now we had been looking forward to the moment we would reach the continual water source of the Oldman River. This river is fed from the Rocky Mountains—now at our backs—and flows east—most notably in our favour!

Oldman River
Pincher Creek to The Grand Forks, Alberta

Dates: June 20 to 27

Route: Oldman River

The Oldman River is a spectacular gorge carving its way through the flat prairies of Alberta. We had no idea such a river existed until we researched it. The Rocky Mountains generate up to ninety per cent of the streamflow; however, the amount of water moving along the river varies from year to year and season to season. This year we were fortunate as the water level was high. The Oldman River basin is the traditional territory of the Piikani First Nation and named after Napi, the old man, who is the creator of the earth and all its living inhabitants. The headwaters of the Oldman River are at Mount Lyall on the border between British Columbia and Alberta. From here, the river flows southeast to the Rocky Mountain foothills and into the Oldman River Reservoir along with the Crowsnest and Castle Rivers. The river then flows east past the Oldman River Dam. The Oldman River continues east across the prairies and is joined by the Bow River at The Grand Forks, which is the confluence of the South Saskatchewan River.

The meandering river offered stunning vistas as we rounded each bend. We gazed in wonder as we paddled past great horned owls, red-tailed hawks, bald eagles and cliff swallows. On our second day on the Oldman River, we rounded a corner and saw what we thought were white boulders scattered along the shoreline—only to find, to our surprise, a colony of American white pelicans, something we never knew existed in Canada. Little did we know that these pure white birds would become a great source of enjoyment for us to watch, all the way to northern Ontario. Pelicans forage by swimming on the surface, dipping their bills to scoop up fish, then raising their bills to drain water and swallow their prey. We often saw them with their beaks tucked back against their retracted necks resting or asleep, on sand or gravel bars. In flight, they are the most impressive. They would fly over our canoe spiralling up from the river valley and soar effortlessly in an updraft. With their wingspan of up to three metres, they were an impressive bird marvellously graceful to watch.

The river was muddy, its fast-flowing current stirring up sediment and silt from below. Here we took water from the river to drink but had to treat it first using our ultraviolet light device. Unfortunately, we could not remove the fine clay sediment so had no choice but to drink it. 'Fibre', we said.

South Saskatchewan River
The Grand Forks, Alberta to Saskatoon, Saskatchewan

Dates: June 28 to July 19

Route: South Saskatchewan River

After seven days of paddling, moving swiftly with the current along the Oldman River, we passed The Grand Forks and entered into the South Saskatchewan River. In Alberta, the river runs for most of its length in a deeply incised valley through some of the most extensive native grasslands remaining in Canada. The South Saskatchewan offers landscapes that are still wild and dramatically beautiful. The river is not difficult to navigate and, except at the City of Medicine Hat, is remote from human presence. Seeing no roads or access points gives one a sense of seclusion. Experiencing an incredible feeling of freedom with this stunning river ahead of us, we take the time to watch and listen; time to gain a sense of history and a sense of place.

The valley of the South Saskatchewan River ranges in width from two hundred to two thousand five hundred metres. In the river valley, you would never know there lies the vast expanse of the open flat prairies above. We stopped along the river and climbed up the high shoreline to take in the views of the canyon below. Carol walked further up into the canyon to see if she could get a glimpse of the prairies to no avail. The views were spectacular as you could see a great distance following the path of the river until it disappeared from view. The South Saskatchewan River is surrounded by the irregular shapes of coulees. Coulees are steep-sided, v-shaped valleys found along the river that once carried meltwater from a glacier. After the last glaciers retreated, the coulees became eroded by water and wind. Coulees are a sanctuary for wildlife and home to hundreds of native plant species.

Before we paddled to the outskirts of Medicine Hat, we came across four paddlers travelling in two separate canoes. The group was having lunch on the shoreline. They were the first canoe trippers we had seen thus far on this leg of the journey. We quickly paddled over to greet them. They said they had started their multi-day trip at the intersection of the Bow River and the Oldman River and were heading to Medicine Hat. Peter, Nadine and their friends were having a short excursion. Together we exchanged stories of our adventures, and they gave us valuable advice on what to expect on the river conditions ahead. They were most curious about our experience and wanted to know what equipment and gear we had. Peter inquired about the composition of the canoe and asked if he could inspect the canoe and the configuration of our equipment. As we approached the water, we were to learn that Peter is a blind person. Peter confidently reached out and asked questions while using his sense of touch to analyze how our vessel was constructed and packed.

Peter and Nadine were no strangers to river adventures and had accomplished many, including paddling a section of the Yukon River. They had also biked from their hometown of Calgary to Ontario on a tandem bicycle and cross-country skied in the Rocky Mountains. We were impressed with their ability to work together as a team, communicating to avoid obstacles along the way. In no way was Peter holding this incredible couple back from enjoying life to the fullest. Lacking sight, his other senses are heightened and give a broader understanding of what most take for granted—sounds, smells and the sense of touch of the world around us. We believe adventurous couples realize the importance of constant communication to overcome anything life throws at them. We also coexist as a team and would not hesitate for a moment to put our lives into each other's hands.

The two couples had equipment more suited than ours to the environment of the river. Wearing rubber boots, they were well-prepared for the muddy walk from the shoreline to solid ground. Because we lacked room to transport footwear for every type of situation, we made do with what we had. We found going barefoot preferable as the mud could suck the sandals right off our feet. Going barefoot and fancy-free was quite stimulating, especially feeling the cool mud squish between our toes.

Along the river consisting mostly of cliffs, we soon discovered the best spots for camping are under Cottonwood trees. Groves of cottonwoods rising majestically in the river valley can live to be over two hundred years old. Whenever we saw an area of cottonwoods towards the end of the day, we were hopeful for a spot to pitch our tent. Because these groves usually are grassy, shady and well protected, we were not the only ones who would enjoy the shelter of the cottonwoods. We had to compete with herds of cattle wanting to drink from the river. The land in Alberta is mostly Crown land under grazing dispositions where cattle wander at will. When we looked for a spot to pitch our tent, we had to be particular to search for an area that did not require a great effort to clean up. On many occasions, the plastic collapsible spare paddle was used as a shovel to toss sunbaked cow patties off the site back into the treeline. Not only did we have to share the land with the cattle, but also the water—the same water which most likely also contained runoff from prairie fields treated with fertilizers and pesticides. Our drinking water was from the river. Even though we treated the water, we were apprehensive of what lasting impact this would have on our health.

Intending to restock our depleted food supplies while camping at nearby Gas City Campground, we headed to Medicine Hat. We paddled into the city. Just under the Trans-Canada Highway bridge, we proceeded to unload our gear from the canoe. We noticed a young couple, Kevin and Kole, sitting nearby at a picnic table and asked for directions to the campground. As it was up a rather steep hill, they kindly offered to drive us and our gear in their truck. We politely declined and proceeded to push and pull our canoe on its cart up the hill to the campground. When we registered for a site, much to our surprise, we learned that Kevin and Kole had already driven there.

They had prepaid a campsite for us for two nights, provided firewood and left a bottle of wine with a beautiful card. We were so touched by this act of kindness and generosity that we were speechless.

The campground staff put us in a designated 'Motorcycle' campsite. The site offered a small paved driveway to park our canoe and lots of grass to spread out and reorganize our food and gear. Our neighbours were two motorcycle riders who had set up camp: Roland with 'his girl' and their friend. They had great fun watching us unload our canoe, wash it, sort our gear and set up camp.

Glenn Reflects:

Carol and I have different thoughts on how to set up a tarp or how to read a compass. Like most couples, we can agree to disagree on how to do certain things. In the end, the task gets done, but it can be a bit of struggle. For example, I read a compass differently than Carol does. But explaining to each other the logic or reasoning behind how we do it is something we would rather not do. We accept the fact that we both do each task differently yet accurately. With a group watching and the skies threatening to rain, I felt under pressure to erect a tarp above the picnic table. Spreading the lightweight tarp out on the ground and visualizing the shape that I thought would offer the best protection, I grabbed a length of rope. I proceeded to head towards a tree only to be intercepted by Carol who suggested a different concept on how this tarp must be erected. As if on cue, Roland, taking this all in from the comfort of his lawn chair, shouted over to us, "Whose idea was it for this adventure anyway?" His comment was a tension breaker and we laughed hilariously at ourselves. We decided moving forward that only one of us would set the tarp up, either Carol or me.

Roland was a colourful, older hippie fellow with a great sense of humour. Roland had a large sports racing bike converted into a chopper-style motorcycle with items that gave him karma and were recycled from home. For example, he had barbell rods for foot pedals, a bicycle bell, beads from his granddaughter and an unapproved helmet with a sticker on it to make it look official in the event the 'cops' stopped him. He wore leather from head to foot—'all cowed up' as he termed it. They were on a road trip across the prairies and, like us, were also seeking protection from an impending thunderstorm. They had erected a giant blue tarp with recycled flexible tent poles. Under the tarp they sat, along with their two tents and picnic table.

It was our good fortune that we had met Peter and Nadine along the shoreline a few days before. Shortly after we had arrived at the campground in Medicine Hat, they contacted us to arrange dinner together. We were delighted to have the opportunity to get to know this couple better. The next day they kindly transported us around Medicine Hat to restock our supplies. They also took us to the Information Centre, where we expected to pick up our food-drop. Before our trip, friends from Kingston had graciously accepted two cardboard boxes which we had packed full of prepared, dehydrated food. Two weeks before our estimated arrival in Medicine Hat, we had contacted them to ask if they would mail one box to the Information Centre in Medicine Hat. It is tough to decipher where and when we would be at any given time as our progress was heavily dependent on weather, wind and undetermined barriers. Unfortunately for us, we arrived in Medicine Hat too soon and our package had not yet arrived. Rather than wait for its arrival, we were eager to resume our paddle excursion so we made arrangements to redirect our package to a location in Saskatoon.

Our Medicine Hat experience was just one example of Canadians rallying behind our cross-Canada canoe trip. Kevin, Kole, Nadine, Peter, the staff at the Information Centre and those who stopped to chat over our canoe parked in the campground all gave freely of their time. They each generously contributed to our trip in unique ways to get us where we needed to go and to accomplish what we needed to do.

Before we left the province of Alberta, we travelled through the Canadian Forces Base Suffield and the Badlands. Suffield is the most extensive Canadian Forces Base and the largest military training base in the Commonwealth. A portion of Suffield is now a national wildlife area zoned for environmental protection. It is one of the largest blocks of predominantly uncultivated grassland remaining in Prairie Canada—roughly half the size of Prince Edward Island. Pronghorn antelope grazed on the side of the river with their young and, in the distance, we saw coyotes eyeing what could be their next meal. The area is stunning. At all times public access to the reserve is strictly controlled. During the military training season, exercises on the Base may require the South Saskatchewan River to be temporarily closed to travel. We contacted the Canadian Forces Base to inform them that we were travelling down the river and passing through their restricted area. We wanted to ensure they would not be firing weapons during a training exercise. They said we could pass through the waterway but strongly recommended not to stop and venture out of our canoe.

The Badlands, we observed, is the most scenic portion of the South Saskatchewan River. Here the valley narrows and the river enters a canyon. Cliffs rise one hundred thirty-seven metres above the river. The Badlands' formations are forever changing. High rates of erosion carve networks of deep and narrow, winding gullies, mixed with hoodoos. The hoodoos are rocks that form weirdly dramatic shapes in a deeply gullied landscape devoid of vegetation except for cactus. Pincushion cactus and prickly pear cactus with their showy, large, colourful flowers are common in this area. They are plants we never knew existed naturally in Canada. This semi-arid region, the Badlands, is greatly affected by short torrential thunderstorms. These storms come with intensity and speed as they sweep across the land. The amount of rain that falls is tremendous, drenching clothing in literally seconds. We experienced this firsthand, as a violent thunderstorm caught us. We had no choice but to escape the river and seek shelter. We erected the tent on the poorly-cemented sandstone and siltstone surface above the river's edge. Just as massive drops started to fall, we quickly entered our tent to wait out the late afternoon storm. With the thunder rolling within the high walls of the canyon, the noise was deafening. Imagine the sound of being trapped in a large aluminum garbage can while someone hits it repeatedly with a hockey stick. While the storm raged, the landscape quickly became a slick and greasy mess of a slow, thin-moving mudslide. We were at a safe distance from the edge and felt relatively confident the tent would not slide off. Watching the rain fall, we could understand how the force of the downpour alone could carve the fantastic shapes of hoodoos all around us. After the storm, we emerged and found it challenging to navigate the slick surface. We were fearful of slipping and twisting an ankle or wrenching a knee. As quickly as the storm had come, it passed, giving us sunshine. Once out of the Badlands, we marvelled at the incredible beauty and wondered why there were no people here. Most, perhaps, rush to the Rocky Mountains and miss all there is to explore in this unique landscape.

One of the sites where we stopped for the night was a large, long sandbar island with a beach at one end and covered in grass. The summer solstice had just passed and we were experiencing long days of daylight. Two hours before sundown, we were just about to fall asleep when we heard a rustle not far away. A coyote travelling down the island must have been startled to find our tent on his patrol. The coyote stood outside on the sand bar and howled at the tent for what seemed to be a long time before realizing we, or rather the tent, was not going to move. The coyote eventually left, with a splash, swimming to the mainland shore. Glenn commented, "It is comforting to know that we have the protection of a fraction of a millimetre of tent material between the coyote and us."

Over many centuries the South Saskatchewan River, twisting and turning to create ever-larger bends, has taken a meandering route. As we crossed the border into the province of Saskatchewan, we found for a short time that we were paddling back into Alberta. Soon after, we turned yet another bend and headed back again into Saskatchewan. The watershed flows east; however, we were never paddling in a straight line; instead, north, south, east and, yes, sometimes west.

The river was relatively deep and fast; however, drought conditions existed the further we headed into July and water levels were diminishing. Past the forks of the South Saskatchewan River and the Red Deer River, the waterway was now more expansive and filled with sandbars. We became skilled looking for the more in-depth sections of water to paddle through as that water is darker in colour. There were sections in the river where we had to walk pulling the fully-loaded canoe as the water was so shallow. Headwinds can be challenging when the river is low and the current accordingly slow. Small choppy waves form with windier conditions making it impossible to distinguish where the channel flows. On this day, a hot prairie headwind was blowing sand in our faces as dunes were now more exposed than ever before.

Carol Reflects:

Being immersed in nature for extended periods, I am connecting more with the environment around me. I feel as though I am forming a stronger spiritual kinship with the land and its creatures. The river to me is a great teacher of humility. It removes non-essentials and the clutter of my sometimes-scattered thoughts. Glenn and I are becoming more in tune with our natural world; we learn to read cloud and wind formations to predict the weather. East winds, with wispy or columnar clouds, bring moisture whereas south winds bring pleasant weather. I am also paying more attention to animals and comprehending their tendencies. We observed a mother duck with her young trailing behind. When mamma duck spots us, rather than head for the shoreline where there is little protection, she guides her flock furiously further down the river. She swims with fluttering precision down the primary current where the river is the deepest. Thankful for our guide, she unknowingly gave us the best route to follow.

People we met would ask, "Where do you camp at night?" There is always a place along a waterway to camp; one has to look for it. We look for a downward slope along a hilly shoreline, a sheltered cove or a patch of grass under a grove of trees. Every morning we start with no plan in place. This way, there is no disappointment or failed expectations—unlike in the working world.

Carol Reflects:

As a child, I spent time camping with my parents and later with my two children. Campgrounds mainly at provincial and national parks are great places for families. They provide a place of discovery and relaxation. Campgrounds have an abundance of people who flock from the cities to get away for the weekend or to enjoy a vacation. I loved campgrounds as a kid. Now, however, my preferences and tolerance levels have changed. I would rather be in a small tent in the wilderness than squeezed together into a small area in a public campground. Being close to other campers usually brings noise and commotion. On this trip, it is a necessity, however, to camp in city campgrounds to resupply. On this day Glenn has his sights set on Eston Riverside Regional Park along the river. As we paddled up to the park, I had an uneasy feeling I could not explain. Perhaps it was the noise of happy campers emanating from the campground. Due to my sense of foreboding, I wanted to go further, but Glenn talked me out of it. During a cross-country canoe trip, one cannot put extra energy into unnecessary tasks.

Camping on the shoreline is convenient as we only have to walk a few steps inland to pitch our tent, set up the tarp and prepare meals. Water for drinking, cooking and cleaning up is close at hand. Communal living brings unnecessary challenges. Campgrounds require campers to set up in designated sites which means we have to assemble the cart, strap in the canoe, load our gear and walk to find the registration office, usually at the other end of the campground at the road entrance. We must pay for a site and then find our site with the assistance of a camp map. Once the tent is up, I go in search of the closest water tap for potable water. Then I find a different place to do dishes and facilities to use the washroom—most of which is usually a distance from our site. I get tired just writing this. I look at the work involved. Glenn looks at it as the opportunity to socialize, have a beer and to make use of the comforts of a toilet. He informs me I am antisocial. So I give in, and we stop at Eston Riverside Regional Park, a regional rural campground complete with all the amenities including a golf course, clubhouse and swimming pool.

It was Saturday, and the campground was full of families, cottagers and lots of crowds. Glenn conferred with the manager who directed us to a seldom-used site close to the shore. Glad to be away from the mainstream and close to the water, Glenn pitched the tent and I set about to make supper. Later, after obtaining coinage from the office to insert into the shower coin box, I took advantage of the facilities to wash the sand out of my hair and dust off my skin. Then, after spending time at the clubhouse socializing with the locals, where golfers collected amongst themselves one hundred dollars towards our cause, we retired for the night. No sooner had my head hit the pillow than we both heard the sound of several carloads of young people. They had arrived at the start of the weekend eager to party. Resigned to the fact that it was going to be noisy, I donned my earplugs.

I closed my eyes, tired after a long day and looking forward to much-needed sleep. Moments later, I sensed a commotion. Glenn later told me that he was startled to see, as he raised his head off the pillow to peer out of the tent, a golf cart careening with wild and reckless abandon across the grass. The golf cart was coming straight for us and at the last minute, veered off. The golf cart ran over the side of the tent and narrowly missed my head by mere centimetres. I was amazed at the speed at which Glenn, clothed only in his silk underwear, bolted from the tent to confront the group. I had never seen Glenn so angry in all the time I have known my husband. A string of harsh words came out of his mouth that would scare the most defiant of humans. There, in his silk underwear, he stood, confronting a group of burly youths who were almost twice his size and a third of his age—and not one of them uttered a word.

No one would accept responsibility for their actions. We were lucky to escape unscathed as the incident could have left us severely injured or worse. Unfortunately for us the tent poles were broken and would need to be replaced. Considering lightweight equipment is costly, this was an expense we certainly could have done without; not only the cost but the inconvenience of arranging future delivery. Glenn reported the incident to the registration office. At the same time, I remained with the tent to deal with a particular cocky youth, the spokesman for the group, who had a dominating attitude. When the seasonal campers heard of the event, they were disappointed and embarrassed that such a thing could have taken place within their community. They informed us we had been placed in the 'party pit'—unintentionally, no doubt. Not being able to sleep in our collapsed tent, we were surprised by a family generously opening their cottage to us for shelter. The next morning, they even temporarily mended our tent poles until we could obtain replacements weeks later. Even though this was an upsetting event, *we were extremely grateful for the kindness bestowed on us, once again, by strangers. The youths never did take responsibility for their actions and quietly slunk away in the night.*

Several days later, it appeared to us that the winds were getting continually more substantial. Whether a windy, blustery day or a labour-intensive portage, these were all determining factors as to how far we moved forward each day. We took one day at a time. Learning by trial and error, we became good at recognizing patterns such as wind direction, wave action and cloud formations. In the last few days we had noticed an increase of cumulonimbus clouds in the distance. These are dense, towering vertical clouds formed from water vapour that is carried upward by powerful air currents. Combined with hot weather, cumulonimbus clouds usually mean some turbulence will soon follow. We had to always be alert as storms might appear more quickly than we could find shelter.

We noticed an increase in recreational boat traffic so we surmised that another community loomed closer. Rounding a bend in the river, we were greeted with a fiercely driving wind forcing us to land. Leaving us with no alternative, the winds had decided it was time for us to get off the water. We steered our boat to shore at Saskatchewan Landing, a sizable recreational area complete with yet another golf course and a campground. Although the "golf cart" incident was still fresh in our minds, we had no choice but to camp at the park. An unnecessary sense of apprehension overcame us. We felt safer in the wilderness with the wild animals than in a campground overrun by people. Reluctantly we removed our gear from the canoe and assembled our cart for the portage. With our equipment in the canoe, we emerged from behind the third hole of the golf course and noticed a party of four golfers preparing to tee-off. Using our proper golf etiquette, we quietly stood and waited as they played through before we cut across the course to the roadway that led to the campground. After arranging a site with park staff and assembling our tent, we set off to do some exploring by foot. We located the park's convenience store and purchased bacon and a dozen eggs for the following morning's breakfast. We only required six eggs for our immediate use so we gave the remaining to the young man behind the checkout. His name was Lundy. A conversation ensued, and we discovered that Lundy was part of the foursome who had played through as we had stood silently and watched. He commented that he had never seen a canoe on a golf course before.

After an uneventful night's sleep, we got up early to beat the wind, had a great breakfast of bacon and eggs and headed to the South Saskatchewan River. A short time later, the fog rolled in, keeping the temperature cool. By mid-morning, the fog had dissipated, and by early afternoon it was hot again. From behind us, we watched a powerboat journey down the middle of the river. It was minimizing the distance between us and soon passed; then stopped about half a kilometre downriver, the occupants casting out several fishing lines. They drifted more closely to our canoe as if they were waiting for us to approach. Surely enough, there was a reason. The occupants of the boat were Lundy's parents and two of their friends, one of whom had also been playing golf with their son. Lundy had told the story of our meeting, and they had come to say hello and wish us well. Much to our delight, they served healthy treats of fresh strawberries, hummus, plantain chips, homemade sticky buns and ice-cold water. We gently floated down the river engaged in conversation, having a wonderful time with these kind folks. They caught and reeled in several fish. We thought briefly of our collapsible fishing rod, securely stored under Glenn's seat in the bow and waiting for the right moment for us to do a little fishing ourselves. And that is all it was—a brief, fleeting thought—as it stayed under the seat for the entire length of our trip. With all good intentions of reeling in fresh, delicious catches during our trip, last winter Glenn had obtained a fishing rod, lures and instructions from a neighbour on the art of fishing. We had had good intentions of taking advantage of remote rivers and lakes to obtain protein for our diet. However, by the end of the day, we were always too tired to attempt the feat of catching and cleaning fish. But we had the rod and the knowhow in case we ran out of food.

Approaching Lake Diefenbaker which is simply a widening of the South Saskatchewan River, we took advantage of our sail to harness the strong wind at our back. We 'flew' fifty-five kilometres in a matter six hours. Glenn hung tightly to the sail as Carol struggled to keep the canoe straight and steady. It was perhaps risky and fool-hardy, but how could we not take advantage of this opportunity? With winds becoming too strong, in addition to seeing approaching cumulonimbus clouds combined with dark, threatening skies, we sought solace in a sheltered cove. As the storm hit with a vengeance, wind, driving rain, thunder and lightning continued well into the night and made us tremble in our sleeping bags. The next day we discovered, through our mobile phone weather application, that the province of Saskatchewan was experiencing a hyperactive week featuring hail the size of tennis balls and at least nine tornado touchdowns. If we had been in a tornado or hail zone, we could have suffered significant damage and possibly not survived. We were relieved that only thunderstorms threatened us.

During periods of rest in the evening, Carol's routine was to compose blog information on her mobile phone to be sent electronically to Rachel at a later date. Rachel would use the information to upload stories onto our website. Since the start of our trip, we had gained many followers on social media. These followers looked forward to our posts. Besides sharing our adventure, they also brought additional awareness for our charity. Most of the time we did not have a cell phone signal except when we were close to a town or a city with cellular towers. As we were nearing Saskatoon, Carol had meticulously prepared several informative emails complete with photos, ready to be sent automatically once her cell phone detected a signal. On this particular day, it was all-for-nought as the emails inadvertently seemed to delete themselves before being transmitted. With battery life drained along with wasted valuable time and effort, this was extremely frustrating for Carol.

Before we reached Saskatoon, we stopped to visit the Berry Barn, conveniently situated on the banks of the South Saskatchewan River. Nestled in a private oasis are orchards where you can pick your berries.

There is a greenhouse, landscaped gardens complete with ponds and a lovely patio with a bakery and gift shop. We took this excellent opportunity to enjoy some home baking as establishments such as this are a rarity along the river. Dusty, with slightly mud-caked feet, we climbed out of our canoe and mingled among the well-dressed tourists and locals. We were searching for a cup of coffee while Glenn decided to sample a slice of Berry Barn's famous Saskatoon Berry pie.

Pressing on to Saskatoon, we were eager to obtain our replacement tent poles and redirected food shipment. We climbed out of the river and portaged to Gordon Howe Campground. After registering for a campsite, we contacted Bill and Dianne, friends who formerly lived in Kingston. The following day Bill and Dianne drove from Regina and not only treated us to a delicious lunch but took us sightseeing and drove us to pick up our supplies. They took us to the Western Development Museum depicting a typical prairie town. The museum enabled us to walk through time. Saskatchewan-inspired stories unfolded as we journeyed from 1910 to the present. Dianne had her upbringing on a prairie family farm, and we learned her firsthand experience of what life was like before farms became modernized to what they are today.

Feeling refreshed and encouraged after a memorable visit with friends, we portaged along the Meewasin Trail and left Saskatoon. The trail passed in front of the downtown core and would eventually give us access to the South Saskatchewan River on the far side of the levee. The levee is constructed of cement and spans across the river controlling the flow of the water as the water cascades over the surface. An object we could not paddle over. The Meewasin Trail is a fine example of an urban walkway as it winds under bridges, through beautifully-landscaped parks and natural areas. Art and benches decorate the walkway. Cyclists, walkers and joggers can gaze at the scenery while dodging a canoe heading for the levee. Once past the levee, we launched the canoe into the water and headed downriver once again. For the next several days, we experienced a fast-flowing current with gentle rapids. We camped on beautiful sand bars every night and gazed at the midnight sky ablaze with the Milky Way. Before daybreak one morning, we woke up to the sound of a deer snorting at the tent. The deer had come down to the water to drink. Being this close to nature is a lovely feeling; life is good.

Approaching Carillon Generating Station on the Ottawa River at Saint-André-d'Argenteuil, Québec.

Camping on Lock No. 5 wharf of the Lachine Canal at The Old Port of Montreal, Québec.

Paddling between the ships on the Saint Lawrence Seaway near The Old Port of Montreal, Québec.

Guiding the canoe through rapids in the fog of the Saint John River, New Brunswick.

Carol prepares a meal using the canoe as a windbreak on Washademoak Lake, New Brunswick.

The Confederation Bridge linking Prince Edward Island and New Brunswick, one of many bridges we paddled under.

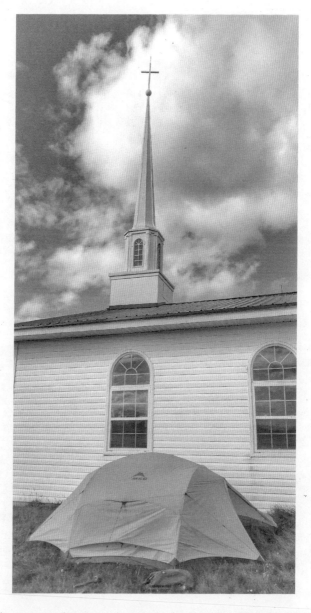

Finding protection from the wind next to the church on Chapel Island First Nation, Nova Scotia.

*Night spent on breakwall beach at the mouth of the
Fraser River in Vancouver, British Columbia.*

Sharing our lunch with Eric next to Crowsnest Highway at Hope, British Columbia.

One of the many tunnels along the rail trail system in British Columbia.

Wild horses turning to look back after galloping past our collapsed tent on the Kettle Valley Rail Trail, Summerland, British Columbia.

The Kootenay River in British Columbia with the Rocky Mountains in the background.

*Top of the Rocky Mountain Great Divide bordering Alberta and British Columbia.
It is all downhill from here!*

Finally on the Oldman River in Alberta.

Typical muddy prairie shoreline, this one along the Oldman River.

*Carol enjoying the shade of a giant Cottonwood on the
South Saskatchewan River near Medicine Hat, Alberta.*

*Taking advantage of what little protection we could find
from the wind on Lake Winnipeg, Manitoba.*

Fire established for cooking and bear/wolf deterrent on Lake Winnipeg, Manitoba.

Carol compares the size of a wolf print on Lake Winnipeg, Manitoba.

Easy lift over with the Turtle Portage Trolley at Lake of the Woods, Ontario.

Paddling amongst the many island mazes through Lake of the Woods, Ontario.

Hoisting up the sail to harness the wind through The Boundary Waters of Ontario.

One of many portages throughout Ontario.

Looking for Indigenous pictographs at Quetico Provincial Park, Ontario.

The spectacular rugged coastline of Lake Superior, Ontario.

Rolling fog banks of Lake Superior, Ontario.

Launching canoe into rough waters using the two-log method on Lake Superior, Ontario.

Waiting for the twenty-one-gun salute at Parliament Hill in Ottawa, Ontario.

Walking canoe through shallow waters on the Mattawa River, Ontario.

Typical lock on the Rideau Canal National Historic Site in Ontario.

Carol's mother and brother Paul escorted us on the last stretch home to Kingston, Ontario.

Saskatchewan River
Saskatoon, Saskatchewan to
Grand Rapids, Manitoba

Dates: July 20 to August 11

Route: Saskatchewan River; Cedar Lake

The South Saskatchewan River meets the North Saskatchewan River at Saskatchewan River Forks approximately forty kilometres east of Prince Albert. At the confluence of the forks, the river becomes the Saskatchewan River. Both source rivers originate from the Rocky Mountains. At Saskatchewan River Forks the current increases and the shoreline becomes inaccessible due to steep banks and the thick growth of dense trees beyond. The river still twists in many directions but now takes a decisive turn north. Since there is no other water source through the prairies besides the Saskatchewan River, we had no choice but to head where the waters flow—straight north. We would be taking the river to Cumberland House, Saskatchewan, the furthest northern point of our journey before leaving Saskatchewan and winding down to The Pas in Manitoba.

Glenn Reflects:

Late one afternoon, we came upon what appeared to be a perfect place to camp: green space next to the river complete with a fire pit. We spotted it as we rounded a bend in the river. The site had level ground with an old windy dirt road emerging from beyond a wooded area. Minutes later a farmer appeared in his four-wheel truck and parked a reasonable distance away from the campsite. He exited his vehicle and, with his fishing rod and small tackle box, walked to the river. Introductions were exchanged as he

117

approached Carol and me, and we asked him for permission to camp on the land. He informed me he knew the landowner. He said he was sure it was okay to camp for the night and to use his name as a reference if needed. Then he went off to catch his dinner. A mere ten minutes later, he returned with three large fish and offered one to us. After a long and tiring day, the last thing I wanted to do was to clean a fish. For someone with experience, it would only take minutes. However, I had never accomplished this feat before, and it would have been a frustrating task at the time, so I declined his kind offer. Envious of his fishing skills, I vowed to myself that someday I would learn the skill.

Settling into the tent around eight o'clock and looking forward to a good night's sleep, we heard loud music resonating from a vehicle as it approached. The vehicle drove right up to our tent and passed it by mere metres before it circled in the grass around us. Once again removing myself from the tent, I approached the vehicle which, I noticed, was a converted van, complete with shag carpet and a massive sound system. Two young gentlemen rolled down the window, and I asked, "Have you come to the water to do some fishing?" They responded by saying, "Hope the noise will not bother you. A group is on the way, and we are going to have a party here. It is going to get loud." My easy-going demeanour quickly changed. After resting both arms on the driver's door, I leaned into the vehicle and replied very slowly and with a voice of authority, "I don't think that is going to happen." I went on to explain our journey; that we had spent a long day paddling sixty kilometres and were exhausted. I suggested they find another place to 'whoop it up.' They quietly left and the party never did materialize. After our golf cart experience, we are leery of sensible youths, acting irresponsibly after being fuelled up on alcohol and driving over us in motorized vehicles.

The Francois Finlay Hydroelectric Dam was the first of many dams on this portion of the Saskatchewan River. In front of the dam, looking for an indication of a portage trail we paddled back and forth from one shore to the other. We were in search of a path or a road to bypass the large fenced-in secure area that encompassed the dam property. While we were discussing how to circumvent the obstacle, much to our delight, a coyote pup came down to the water playfully seeking our attention and amusing us with his acrobatic stunts. After several minutes of being entertained by the coyote's antics, we took our discussion back to the task at hand. After considerable head-scratching, trying to figure out how to get around the darn dam, fortunately, we located a phone number. The number put us directly in contact with the control room operator. The operator, sitting in his room, was viewing closed-circuit cameras and finally spotted us in our canoe. Continuing to watch, he gave instructions as to where we should paddle. He and his staff would meet us and assist in our portage he said. He made it very clear their assistance was not to place us back into the water but to bring us to a location where we could safely reach the river on our own. With canoe and gear, through multiple, locked, fenced gates, we were transported to a roadway beyond the dam. The control room operator gave us some valuable information. He said, according to their safety policies, it was the responsibility of SaskPower to provide assistance to paddlers to safely cross any dam in the province. We found that hydroelectric dams, all the way to Lake Winnipeg, Manitoba, were similarly confusing and a challenge to circumvent. Therefore, for every hydroelectric dam we had to bypass in the months ahead, we would call the control room operator and introduce ourselves. We would explain the nature of our journey then ask them to kindly adhere to their safety policies to assist us safely around their obstruction. This always produced favourable results.

On our portage from the other side of the dam, we witnessed a tremendous number of trees and brush piled high on the land. The melting of the Rocky Mountain snowpack, combined with unusually hot temperatures, caused torrents of water to rush down the North Saskatchewan River virtually razing islands clear of vegetation. The evidence of the devastation was here at the dam. Dam authorities used large equipment to remove the log jams to prevent the debris from entering the turbines.

Four hundred fifty kilometres northeast of Saskatoon lies the remote village of Cumberland House. Before reaching Cumberland

House, the last community before reaching Manitoba, we happened to pass a group of buildings and assumed it was a fishing lodge. Seeing any trace of buildings or homes this far north along the river is extraordinarily rare. Glenn called out and shouted. "Hello, is anybody there?" There was no answer but several large friendly dogs and a beagle, wagging their tails and wading out into the shallow water, came out to greet us. We got out of our canoe to greet the dogs, but they were so excited to have visitors they almost knocked us down in the process. We were quick getting back into the canoe to spare being scratched by dog nails and getting covered in river mud. We continued to paddle downriver, disappointed we did not have an opportunity to chat with a local. Looking back over our shoulder, we saw the beagle had left the other dogs and had decided to follow us. Although we called out to the beagle to 'go home', he seemed persistent, trekking along the shoreline with his short legs and leaping over logs and rocks to keep up with us. Forty minutes later, with the beagle still pursuing us, we spotted a black bear foraging for food just ahead on the shore. It appeared the beagle and the bear would be on a collision course. Carol spoke first and commented, "This could be interesting." Neither animal was aware of the other's presence. From a safe distance in our canoe, we watched the interaction unfold. The beagle approached, suddenly stopped dead in his tracks, having spotted the bear approximately five metres ahead. The beagle threw back its head and bayed a low moaning howl that only hounds make. Startled, the bear stood up to get a better look. Then both animals turned in opposite directions, taking flight and not wanting to confront one another. The whole scene was a bit comical.

The land along the Saskatchewan River was transforming before our eyes. Gone were the forests and the rocky shoreline of firm solid ground. Gone were the glimpses of lush farmland, the fields of yellow canola and the herds of cattle we had been familiar seeing near Saskatoon. The river was quickly becoming an incomprehensible maze. We had entered the delta. Where poplar and willow once inhabited the muddy shoreline, now is replaced by tall, dense, man-high cane grass. We were inexperienced paddling in this unfamiliar terrain. We had only our

GPS to rely on—and depend on it we did. Glenn did an excellent job of guiding us through the maze of tributaries. At ten thousand square kilometres, the marshes surrounding Cumberland House make up one of the most biologically diverse places in Canada. One can easily get lost in the maze of wetlands. Sadly, many dams we came across in our journey impact the environment and play havoc with the ecosystems. The flow of $H2O$ that gives the nutrients of life to the marsh and the animals is gradually diminishing. The dams divert water for lawns in southern cities and expand irrigation and industry. The area is affecting the livelihood of the people of Cumberland House, stewards of the Delta. Hydroelectric dams are killing downstream wildlife by turning natural flow patterns upside down. A free-running river has its most massive flows in spring and summer. A hydroelectric dam stores this vital summer water in the reservoir then dumps it through the turbines in the winter when electricity demand is highest. Not only has it affected river fish and animals but moose that once were abundant in this area are drastically reduced.

Following the current was an indication of the river's main path. It could become confusing at a fork in a tributary that often branched out in several directions. We took some wrong turns, and the most notable error was going down an arm and completely missing the community of Cumberland House. Inadvertently bypassing a critical National Historic Site of Canada was rather disappointing. Cumberland House was the Hudson's Bay Company's first inland fur-trading post and Saskatchewan's oldest permanent settlement, founded in 1774. More importantly to us, was missing the opportunity of visiting the Cree First Nations community. Over the winter we had contacted Grand Chiefs of several First Nations informing them of our journey. In particular, we wanted to let them know we were passing through their territories and that we were building awareness surrounding food security. Twelve per cent of Canadians live with food insecurity; for children and adults living on and off First Nations reserves, however, it is as high as forty-eight per cent. Regardless of the area's diminishing habitat, it is a goal of First Nations to strengthen and renew the indigenous relationship to the land and to traditional practices of harvesting, hunting, fishing, preparing and preserving food. Transferring the knowledge about traditional methods from elders to children and youth instills increased pride in the indigenous identity and restores the essential connections between culture and healthy living.

Having spent all day in the delta, which encompasses the Saskatchewan River, we managed to find an uncomfortable place to camp. We were thankful to find anything as the possibility of sleeping in our canoe had crossed our minds. Our coordinates were just south of the Cree First Nations community. We found a small patch of rocky, uneven land surrounded by marsh and millions of mosquitoes. Hordes of mosquitoes and black flies are notoriously common in many parts of Canada. We donned our bug jackets, sprayed our clothing with insect repellent and set about doing camp chores. Having a taste of what it is like to have swarms of insects invading our space made us thankful. It was not something we had to put up with for weeks at a time as is usually the case. Before dusk, we could hear the sound of mosquitoes making their way down slowly from high in the tree canopy. We quickly dove into our tent before the ambush. As we lay in the protection of our tent, the onslaught of insects was a sound like no other. It is not that we are unfamiliar with the hum of mosquitoes; however, the combined sound of millions of the little pests was similar to the distant droning of a helicopter. As we watched mosquitoes slowly accumulate on the inner screen of our tent, we were thankful we were not spending the night in an open canoe.

The people at The Pas in Manitoba knew we were arriving, we guessed, by the way people approached us and asked, "Are you two the paddlers?" Word of our upcoming arrival must have been carried forward by the few who had spotted us along the way. Locals approached and offered information on what we could expect as we advanced through unfamiliar terrain. They invited us to use their camps along the way and talked about local indigenous history and customs. We were now forty kilometres from the border of Saskatchewan and felt relaxed in The Pas. We spent a few days to recharge and rest. The Pas was the last opportunity to restock our food and fuel supplies before the long paddle to Cedar Lake and down the coast of Lake Winnipeg. We needed enough supplies to last for three to four weeks. With calm conditions we estimated that we could paddle the distance in less than four weeks; however, one could be windbound for days at a time on either lake.

Food is expensive in The Pas. We paid twice as much compared to what we usually do. Food and supplies at indigenous communities further north are even more costly, leaving people to travel by train to The Pas to buy supplies. Once a month, people come and spend several days shopping; filling up boxcars of food for the return trip home. We spoke to one individual from a northern community who commented that he pays thirty dollars for a five-kilogram bag of potatoes compared to our five dollars' cost. One litre of milk costs twelve dollars compared to our four dollars' cost.

Leaving The Pas, we continued down the Saskatchewan River towards Lake Winnipeg. This region encompasses the southern expanse of the Saskatchewan River freshwater delta. As we had in the northern area of the delta, at times, we became disorientated. The river, which was nothing more than a ribbon cutting through floating grasslands, had disappeared completely, revealing the great expanse of Cedar Lake. Our GPS showed us still paddling on the river, but our eyes told us otherwise. After several long moments of bewilderment, we realized that high wind and wave action perhaps had dislodged the islands of grasses dispersing them somewhere into the vastness of the lake. Thankfully, the water was relatively calm for the next couple of days, and we made our way to the east end of Cedar Lake. Now close to Grand Rapids and the entrance of Lake Winnipeg, we secured the only piece of accessible shoreline in sight. A bare two metres existed between thick forest and the water's edge. Here we pitched our tent on the exposed, dramatically-slanted, rocky shoreline. Thankfully, the weather remained calm and we did not have to deal with crashing waves. While in the delta, First Nations community elders graciously extended an invitation for us to camp freely at their hunting camps or

Pow Wow grounds. They knew that we were up against having virtually no other solid land available to pitch a tent. Not only were we grateful for these opportunities, but it gave us insight into the genuine kindness and hospitality of our First Nations' peoples and their willingness to share their lands.

Lake Winnipeg
Grand Rapids to Traverse Bay, Manitoba

Dates: August 12 to August 31

Route: Lake Winnipeg

From the headwaters of the Rocky Mountains along the Oldman River to Cedar Lake, we had completed paddling our longest continuous river thus far, just under two thousand kilometres. The Saskatchewan River ends at Grand Rapids and, for us, the beginning of Lake Winnipeg.

Many rivers feed into Lake Winnipeg: the Saskatchewan River, the Red River from the south as well as the Winnipeg River. The Winnipeg River drains water from Lake of the Woods and Rainy Lake from the southeast. Lake Winnipeg empties northeast along the Nelson River into Hudson Bay.

On our second day on Lake Winnipeg, we were approached by a group of local fishermen tending to their nets. They stopped and inquired where we were going in our canoe. It was an excellent opportunity to gather information about the upcoming shoreline. We enjoyed their humour and the gift of a fresh fillet of fish ready to be pan-fried. They gave us two pieces of invaluable advice. Firstly, if the wind comes from the northeast, get deep into a cove and be prepared to stay there for two to three days. A northeast wind comes from the far shore, gathers momentum and releases all its fury on the western shoreline of the lake; exactly where we were paddling. The second was a strong caution given about upcoming Long Point. It is a continuation of the glacial feature known as The Pas Moraine that extends fifty kilometres into Lake Winnipeg. We were soon to find out for ourselves how unforgiving Lake Winnipeg can be.

The elements could be totally out of our control and, very quickly, one could run into a dangerous situation. This day we would capsize. It was not our proudest moment, but we had always anticipated it could be a possibility. Lake Winnipeg is a vast, shallow, unpredictable, 'hot mess' of a lake which demands respect. Wind causes waves. The size and shape of a wave depend on how big the body of water is and its depth. For example, Lake Superior's waves resemble an ocean swell which we found easier to paddle in than on shallow lakes such as Lake Winnipeg. Shallow lakes have waves that are steeper, closer together and break more easily. They are difficult to paddle in.

We chose to paddle down the western shore of Lake Winnipeg as wind generally comes from the west. It was best to stay along the lee shore as waves in the middle of the lake are larger. The day we capsized was a long and harrowing event for us. Having arisen well before daybreak, we set out to take advantage of the early morning calm. We made our way onto the water in the darkness just after four a.m. We were travelling south, but Long Point took us east—approximately fifty kilometres—close to the middle of this dangerous shallow lake. To our dismay, large boulders and rocks cover the shoreline. There was absolutely no opportunity to come ashore during strong surf without doing significant damage to our canoe. It was eleven in the morning, and we were still not able to stop for a much-needed break. Without having had breakfast or an opportunity to even take a drink of water due to the increasing challenge, we faced paddling in strong wind and waves. Having rounded the point, we now attempted to head west back to the south shoreline. We were committed to continuing, as there was no other place to go. Five hours of paddling had passed from the time we had rounded Long Point. With increasingly rough water, cresting waves and suffering from exhaustion, we had spotted a beach and made our way towards it; to no avail.

We had been caught off guard by a massive wave that hit us broadside and immediately flipped us over. Once we were free of the canoe and had surfaced, we were relieved to see each other's head above the water. With paddles clutched in our hands, we needed to swim to shore and bring the canoe with us. Our flexible waterproof canoe cover with an opening at each end to sit prevented our gear from escaping. This spraydeck we had had specifically fitted to our canoe to repel rain and waves but, in this instance, it served a purpose to contain all our gear. Once we made it to shore, we were in a mild state of shock. We realized the situation could have been much worse. We spread out the articles that got wet. Fortunately, most items remained dry due to our waterproof bags. Relieved to be off the water after twelve hours of strenuous paddling we thankfully set up camp for the night.

Due to the isolation, we had experienced on the lake so far, we were amazed when we saw, walking down the beach towards us, two First Nations men with rifles slung over their shoulders. The men informed us that they were on a family weekend retreat camping high above on the cliff. They had witnessed us disappearing and reappearing behind each immense wave in the high swells, as we approached the beach. Having lost sight of us, they had come to investigate. They told us to prepare a big fire for the night as bears and wolves would frequently visit this beach. We took their advice and collected all the driftwood available. Glenn built the most massive fire you have ever seen! Sleep this night came quickly for us, exhausted as we were. The next morning two others from the same family high on the cliff came down to make sure we were still alright. They inquired as to when we would be leaving and informed us of what to expect going down the coast in the way of endless beaches and small coves. Then they headed off, leaving us to have breakfast and pack up our gear. A short time later, we departed once again into the surf to continue our paddle down the coast. To our amazement, we found the entire family had come down to the shoreline to see us off. They waved and shouted their encouragement to us! We felt humbled by their presence. It was an authentic gift of kindness and an outpouring of support that left us with a feeling of love for our fellow man. We did not feel as isolated anymore. These were the last people we were to see for the next several weeks.

Because of the remote areas we were travelling through, and considering the incident we had just experienced, we required a device to summon help. For our peace of mind and, in the event of an emergency, we had brought a SPOT communication device that uses a GPS to determine our location every sixty minutes. The SPOT uses the Globalstar Satellite Network to transmit those GPS coordinates to a map located on our website for anyone to track and follow us. The device also allows us to notify friends and family of our exact GPS position and status on a consistent, routine basis. In the event of an emergency, the device will allow us to send those GPS coordinates to emergency rescue personnel.

"Lake Winnipeg's mood swings are those of a temperamental, but an alluring, goddess," wrote author Paul Sigurdson. He could not have portrayed the lake more accurately. Some days the lake was incredibly calm with the intense sun shimmering on the water. At the same time, the nights were just as peaceful, brilliantly lit by the stars of the Milky Way. It was as if the goddess of the lake were asleep and at peace. Then there were evenings where the winds change in a subtle manner and the next day the fury of the fiercely independent goddess was upon us. The wind could rise at any time without warning. It is then that paddlers and freighters await the lake's fury. Lake Winnipeg's northern basin is an immense expanse of water, a strange and mysterious place. If we had not been careful, we could have been lost in the vast, trackless waste of islands, sandbars and bays. The surface of the lake is like an artist's palette with its own unique style. The lake pushes water around basins and between its islands. Below the surface of the water there is an entirely different landscape. Lake Winnipeg is an unpredictable force of nature that can lull you to sleep one minute and, the next, spit you right out of your canoe.

Storms come quickly, and the waves break hard cresting with overhead thunder and lightning streaking through the dark sky. The coastal shores are abundant with some of Canada's whitest and best beaches on which to camp. But these beaches lack protection from the elements. As beautiful and as wild as the coastline is, accessing the swamps and forest behind the beaches was not an option for us due to the thick tangle of foliage.

Glenn Reflects:

Several times we were windbound on Lake Winnipeg. I would often joke and sing the theme song from an old television show, 'Gilligan's Island'. I told Carol I felt like a castaway. I wondered aloud if we were ever going to get off Lake Winnipeg. On one particular windbound evening, I wandered down the beach to clean the dinner pots. The sun was setting, casting long shadows: A stiff wind was blowing into my face as waves crashed down, disturbing the sand. I heard something sounding like a voice. I turned to look only to find the vastness of my surroundings. Surmising it was only the wind singing through the trees, I continued a few steps only to hear it again. Once again, I heard a voice speaking to me. I turned without seeing anyone except for Carol, who was too far away and not even looking in my direction. I stood in bewilderment. I could not distinguish what I was hearing, so I closed my eyes to focus while trying to eliminate the sound of the crashing of the waves. I convinced myself it was the wind; we were isolated and had been for many days. With the sun at my back, bent over scrubbing away at the pots, I was startled by a shadow looming over me. My heart leapt into my throat as I quickly straightened up and turned around. Carol had brought me a dish I had forgotten! "You scared the crap out of me—I thought you were a bear!" Carol stared at me; I explained the voice I had heard. No sooner had the words left my mouth, when glancing past her, I saw a large black bear lumbering down the beach towards us. "Here comes a bear," I say. Carol, looking directly at me, smiles and says, "Nice try." As the bear approached, I jumped up and down while frantically waving a spatula, yelling, "Whoa, bear!", "Whoa, bear!"

This large black bear is not aware of our presence. He had not detected our scent nor heard my warning as he continued his advance. Carol calmly states, "Try banging the pots together—he cannot hear you." Once the bear had heard the banging of the pots, he stopped, looked in our direction and stood up on his hind legs with his nostrils flared. Recognizing we were there, the bear turned around and ran away with incredible speed. We were close enough to see the hair on his back bristle, and we felt terrible for scaring the poor animal. That same night I built the most immense fortress of driftwood around our tent one could imagine—not to mention a roaring fire nearby. But before I closed my eyes, I could not help but wonder if the voice I had heard was a spirit warning me.

Days later, happy to be on the water after three nights of being windbound, we were paddling along the shoreline looking for a camp-site. A beach caught our attention. It was complete with a small river entering the lake which seemed like a prime spot to camp for the night. As we approached, Carol spotted a wolf venturing towards the beach. The wolf appeared to be coming out of the thicket towards the mouth of the river. Moments later, the wolf was joined by others until a large pack revealed themselves. Like the bear, the pack did not sense us due to the wind direction and noise of the surf. We stayed as still as possible as we sat and watched these magnificent creatures scour the shoreline. It is unusual to witness wolves, and we relished the moment with wonderment and awe. "Well, Glenn, let us just leave this beach to the wolves," Carol said, not wanting to encroach on their space. With no argument from Glenn, our combined decision to relocate was an easy one as we paddled further down the coast to choose another beach less inhabited.

If the goddess of Lake Winnipeg was in a tranquil mood, we uti-lized these days to gain as much distance as possible. Each calm day we rose early and finished late. We put in long days of paddling until our bodies would grow weary. Much like we had experienced in the Northumberland Strait, we were jumping from point to point across bays and making valuable distance. We noticed the air quality changing around 'Wicked Point', a cape two hundred and twenty-two metres in height, that extends into the lake. On calm days the sky was filled with a haze-like fog blocking out the sun. As the haze became thicker it became apparent the air was dense with smoke, we assumed, from a forest fire. The situation we were now in only added to our nervousness and feeling of isolation from civilization. Not only were we a week's paddle from any community but we had no service on our marine radio or our mobile phone. With no communication, we were not sure about the location of the fire. We had to use our GPS as we could no longer see the shoreline. We covered our mouth and nose with balaclavas to screen out the smell of smoke. We found out much later in the First Nations community of Jackhead that forest fires were raging in British Columbia. Due to wildfires that were burning out of control and devastating more areas than any other season on record, the province of British Columbia had declared a state of emergency. The smoke from these fires was experienced not only all across Canada, but as far away as Ireland.

When the winds changed from the west to the south, the smoke went away. It seemed the goddess of the lake had given us one challenge after another. We were now on the lookout for coves in which to seek shelter almost every night. On one such day seeking the best protection possible from the force of the wind, we were forced into a small sandy cove only to find out that the next day the wind direction had changed. Returning to the water to relocate was not an option. We packed up our gear, secured our canoe to a tree and walked down the shoreline around several corners to find a cove that was better protected. Relocating our campsite would entail making several trips back and forth but what else did we have to do?

Taking advantage of being windbound, we set about doing camp chores that we usually did not have time to do. We organized the food barrels to take stock of provisions. We washed clothes, mended worn clothing and equipment, and relaxed and read. Glenn discovered several cracked ribs in the canoe and spent time repairing the Kevlar with some good old duct tape. To preserve precious fuel, we cooked our meals over the fire rather than use our ultra-light camp stove. We would dig a shallow hole in the sand, place two logs over the hole to act as a grill and build a fire underneath. Until now, all across Canada, we had cooked over our stove rather than make a fire. Due to dry conditions experienced thus far, we did not want to take a chance of starting a forest fire. Plus, we practised no-trace camping as a rule.

Later in the day, we went for a walk along the coast to beachcomb, explore and to stretch our legs. While staring out at the white caps on the lake as far as the eye could see, we wondered once again how many days we would be windbound? Returning from our hike, we spotted a bear emerging from the forest. It noticed us after several moments and turned heading from whence it came. We both looked at each other, not saying a word. Another massive fire this evening was in order.

Carol Reflects:

Lake Winnipeg is so unusually different from any lake I have ever paddled before. After spending two days windbound on a sandy beach, we set out, anxious and eager to make some distance. We decided to take advantage of a lee westerly wind, paddle close to land, yet stay a comfortable distance from the shoreline. Mesmerized by the coast and lost in the thought, I found that we had become stuck on a sandbar. Noticing that I had inadvertently paddled into an area that was land-locked by an unforeseen rock shoal, I got out of the canoe to give us a push. I put one foot out of the canoe, lost my balance and fell into the water. I had my waterproof waders on for protection from the chill and the waves. My waders immediately filled with water soaking me from chest to toes. Worse yet, I now had a substantial amount of water in my wader boots.

I wondered what would happen if I ever fell overboard. Would I sink like a rock? It was fruitless to drag the canoe over the rock shoal, so we backtracked for over an hour, which was valuable time lost. Usually, this would not bother me, but I was having a bit of an off day today. Paddling on big water with the canoe rising up and down in the waves gave me motion sickness and made me dizzy. Not an hour later, the mystery of the lake baffled me once again. This time a long sand bar, also an extension of unforeseen land, had locked us in. Once again, I got out of the canoe, lost my balance and fell into the water. At that point, the weeks of physical exhaustion, the frustration of not being able to stand on my own two feet—not to mention the mental stress of paddling on this unpredictable lake—got the best of me. I just sat in the stern of the canoe and had a little cry. During moments like this, it is best to go to shore, take a long break, recharge and refocus. Similar to life situations that tend to overwhelm, frustrate or discourage me, I do my best to adopt the same strategy—pause to decide another course of action. There is always an answer to life's dilemmas; one only has to take time to take a break to do some inner searching.

We were nearing Jackhead, a remote community of two hundred Kinonjeoshtegon First Nation peoples. Having left the peoples of the Cree Nations further north, we were entering into the lands of the Chippewa (or Ojibwe) peoples. With August drawing to a close, the high winds of September were encroaching. There was fear in our bellies as we worried about our safety paddling through 'The Narrows', which lay before us. We were worried about what the goddess had in store for us as we had to paddle through a narrow channel separating the northern and southern basins of Lake Winnipeg. Our plan was to seek guidance from the elders at Jackhead.

Once we reached Jackhead, we landed on the beach and set up camp in the treeline. We chatted with a few locals and they told us paddlers and visitors to their area were rare. Glenn was off finding firewood and Carol was about to make supper. As she started cooking, the scent of delicious, homemade, dehydrated, gourmet meals aroused the senses of the local domestic dogs from the village. Carol, engrossed in her work, did not notice as a pack of dogs, quietly ready to snag something to eat, had surrounded her. Glenn, the alpha male, approached in the nick of time, scaring them off. Luckily, they obeyed his stern commands and the dogs left as quickly as they had come.

The next morning we headed into town to find the Chief. We felt safe leaving our canoe and gear unattended on the beach. Our intention from the start had been to travel across Canada solely on our own steam whether it was portaging, pushing our canoe or biking. Now we were in a predicament as we did not feel safe paddling through 'The Narrows'. We planned to get direction on what our best course of action would be at this point. We met with the Chief at the village office. He understood our dilemma and kindly offered to transport us, our canoe and gear to the Municipality of Bifrost-Riverton, bypassing the hazard of The Narrows. We were grateful for his kindness and his assistance. He gathered members of the village together and they sprang into action in their willingness to help two stranded paddlers. The Chief's daughter promised to take time off from her research the following day to drive us to Riverton. She was on a government contract to research the effects of climate change on Lake Winnipeg. The Chief's daughter was a lovely, warm and wonderful person and got us safely to Riverton. We vowed that someday in the future we would return to paddle the section of the lake that we had missed. Saying our goodbyes to the Chief's daughter, we thanked her for her kindness and continued down the west coast of the southern basin of Lake Winnipeg.

Days later we reached Gimli, the small scenic town founded by a large group of Icelandic settlers who arrived in the 1870s. Gimli is a pick-up for our food-drop that had been mailed to Evergreen Basic Needs. Back home, our friends, solely, had organized this food-drop. Evergreen Basic Needs is a charitable, community-based organization working to eliminate hunger and food insecurity through food distribution, education and support programs. The organization also has a delightful café, a second-hand store and provides valuable support for its community. Our friends could not have picked a more fitting connection for us. Executive Director Karen was instrumental in helping us along with this transition. Karen, and her husband Lee, invited us for dinner. She also arranged for her neighbour, a news reporter with the Express Weekly News, to publish a story.

Straight across from Gimli was the Winnipeg River, our next water source. We could almost envision the other side of this vast lake and contemplate the risk of travelling twenty-five kilometres straight across. The other option would be to paddle approximately seventy kilometres around the southern basin of Lake Winnipeg to come up the other side. Considering the ever-changing moods of the lake and after experiencing a thunderstorm, we took it as a sign to take the safer route and paddle around the coast of the southern end of Lake Winnipeg.

Paddling around the coastline, we passed the mouth of the Red River. The water is shallow and filled with silt and sand bars. Paddling was strenuous until well past the opening as the canoe was continually running aground. As with any river flow, the current picks up sediment from eroding shoreline and forms deposits of sand bars at the mouth. Not only has the area been affected by river sediment but also a known phenomenon called a 'setup' or 'wind tide'. On long, relatively-narrow lakes such as Lake Winnipeg fascinating wind and wave effects occasionally take place. When northerly winds blow along the length of Lake Winnipeg, they exert horizontal stress on its surface. Surface waters move in the direction of the wind and pile up along the windward south shores. This is called a setup. Records show along many beaches on southern Lake Winnipeg setups greater than one metre above normal lake levels have occurred. The associated high waves with their uprush effects have caused considerable storm damage, backshore flooding and shoreline erosion. The highest setups occur in the fall when the northerly winds are most potent. If the winds die down suddenly, the waters rush northward and then slosh back and forth in a process called seiching.

Pitching our tent on a beautiful large sand bar surrounded by a slim band of poplar trees, we camped cautiously along the southern shore of Lake Winnipeg. With Lake Winnipeg to the north and the Red River to the south, it was a spectacular location indeed. Of course, a north wind picked up early the following morning forcing us to set off before sun-up. Heaven forbid if we had become stranded on this sand bar with absolutely no protection and the seiche phenomenon had occurred. With perseverance, we made it to a public dock and had to walk/portage the rest of the way to Grand Beach Provincial Park close to the mouth of the Winnipeg River.

Winnipeg River, Lake of the Woods and Rainy River
Traverse Bay, Manitoba to Fort Frances, Ontario

Dates: September 1 to September 29

Route: Winnipeg River; Lake of the Woods; Rainy River

Now at Traverse Bay at the mouth of the Winnipeg River, we were happy to be past big, bad Lake Winnipeg. Out of twenty days on the lake, we had spent eight stranded, windbound for more than one day at a time. Moving forward, we realized now several other factors would be a challenge for us. We were paddling against the current of the Winnipeg River. Before reaching the Manitoba/Ontario border, we would be required to circumvent six hydroelectric dams. With September upon us we wondered just how much farther we could travel before the canoe paddling season would end.

We came to discover that bears love to feast on protein-rich acorns to bulk up for the long winter hibernation. We found this out after we had selected an ideal spot to pitch our tent at Whitemouth Falls Provincial Park where the Whitemouth River empties into the Winnipeg River. The day park was a treat to find at the end of the day just before we would be skirting another dam. The park's forest includes sizeable mature oak trees where bears regularly gather to fill their bellies with thousands of delicious nuts. At this time of year locals, from the safety of their cars, would gather to witness the bears feeding. Several people approached us and expressed concern that we were camping in the area under the oaks. We informed them we had no other place to camp. We rather liked the spot and we assured them we would be diligent with our food storage. We had a restless evening.

We tried to sleep with one ear open as we listened to the bears grunting and foraging as they vacuumed up the acorns that littered the ground. Bears were still roaming in the morning. We became comfortable, yet guarded, with the bears being around. When a bear got too close to the tent Glenn said, with a voice of authority, "Okay, go find another tree. Off you go! Now!" Bears were more interested in eating their nuts rather than in us or our food. Both parties kept a wary eye on each other as we each consumed our separate meals.

Several days later we crossed over the border into the last province on our route, our home province of Ontario. We took a few minutes to celebrate. Even though we still had three thousand kilometres to reach the end, it felt sort of close. At this point, we were celebrating everything. The water seemed more transparent than the waters located in the prairies. With large areas of exposed igneous rock, the terrain looks more like the Canadian Shield. The sky was blue and the weather remained better than expected. Our goal was to reach Fort Frances, which is at the headwaters of the Rainy River, before the season closed in on us. The Rainy River, a narrow river over one hundred and thirty kilometres in length, drains from Rainy Lake into Lake of the Woods along the Canada/United States border. If we attempted to stop before Fort Frances the chances of paddling against the current of the Rainy River during next year's spring thaw would be impossible. Carol was determined and persistent in her view that getting to Rainy River would be possible as long as the weather cooperated. However, Glenn wanted to call it a year as his muscles and joints had begun to feel the effects of paddling. With Carol's positive encouragement, we decided to push forward.

Northern Ontario is frequented by many Americans who arrive to join local fishing guides in hopes of landing the big one. Numerous times we were passed by small motorboats transporting customers from one fishing hole to another. One boat with friendly tourists and a very knowledgeable guide stopped us. They handed over the leftovers from their shore lunch, fresh pan-fried pickerel. The group asked in a southern drawl if we had seen any 'bars' on our journey. Because of puzzled expressions on our faces about what 'bars' meant, the guide diplomatically reworded the question for us. We replied, 'Yes, lots of bears.'

The Winnipeg River continues to travel past the communities of Whitedog, Minaki and, finally, to the City of Kenora where it connects to Lake of the Woods. Viewing Lake of the Woods on a map seemed overwhelming with the lake's maze of over fourteen thousand, five hundred islands. As we paddled, we felt like we had become lost in a sea of islands. Using our GPS, we meticulously plotted the original route of the voyageurs who travelled this waterway as they transported furs for commerce before this country became established. The route would take us between islands of beautiful red pine trees and vaulted rock formations.

We had a rare opportunity to experience a less traditional method of portaging over land. Just south of the Township of Sioux Narrows, past the pristine waters of Whitefish Bay in Lake of the Woods, we came upon Turtle Lake Canoe Portage. It was a hand-powered trolley that allowed boaters to move their small vessels overland between two bays. It eliminates the one hundred forty-five-kilometre trip around a peninsula. After carefully reading the Trolley Operations and Safety Rules provided by the Ministry of Natural Resources, we paddled the fully-loaded canoe right onto a partially submerged trolley. Once the canoe had become balanced on the trolley, we cranked the sizable red wheel to quickly and easily draw the trolley up and over to the other side. About five minutes and thirty metres later, our canoe slid smoothly into the swampy waters of Turtle Lake. So amusing to have this unique portaging experience!

The days were beginning to get colder and the skies more threatening with rain. Carol was encouraging Glenn not to quit for the year. She reminded him of the progress we had made and that, for the third week into September, the weather could be a whole lot worse. So with renewed determination, we continued to Fort Frances, where we planned to finish for this year.

We were fortunate to meet Karen at a local campground. Karen worked at the campground and lived down the coast toward the Assabaska Ojibway Heritage Park. Karen is the type of person whom, hopefully, we all have had the opportunity to meet at one time or another—a person you instantly feel like you have a connection with and have the sense of having known her all your life. She extended a sincere and heartfelt invitation to us to stay at her home on our way through. We humbly took advantage of this kind offer as the thoughts of a warm meal, hot shower, and a warm bed were exactly what we needed on this cold night.

Carol Reflects:

The kindness of strangers we met is always and, will always, be a humbling experience for me. Canadians were so genuine in their complete openness and interest in our travels. The hospitality we received was given generously and freely. Those who gave it wanted nothing in return. Karen and her husband Tony welcoming us into their home with complete trust and sincerity is a good example. We had nothing but stories of our travels to give in return. It is the gift of meeting fine people like Tony and Karen and the many others before them that instilled the pride I have in this country we call Canada. I hope that someday they will have a glimpse into the depth of what their gift meant to us: to read this and understand the importance of their actions to a cross-country paddler. The string of human connection given by people like Tony and Karen not only binds us together as Canadians but, truly, is a statement of the kindness of the human spirit.

Exiting Lake of the Woods and heading towards the mouth of the Rainy River, we navigated towards Windy Point, an island on Lake of the Woods. Once again, we were forewarned by locals to be careful. Windy Point, they said, left paddlers dangerously unprotected and fully exposed to south-west winds coming from the lake. Digging our paddles into the water and heading straight into substantial, crashing waves, we attempted to round the island of Windy Point. Waves were continually breaking over the bow of the canoe and soaking us with cold water. Paddling like the hounds of hell were at our heels, we managed to round the Point using every ounce of strength we had. Now soaking wet, we managed to find a protected bay and immediately changed into dry clothing. With the temperature only being close to five degrees Celsius, we needed to warm up instantly to risk exposure. Glenn was so cold he was shaking uncontrollably and, swathed in his down sleeping bag, he sought warmth in the tent. Carol donned every piece of clothing she had to stay warm. She set about doing camp chores and prepared a quick dinner, followed by the necessary cleanup.

Along the border of Minnesota and Ontario is the Rainy River. It certainly lives up to its name. With increased rainfall, mixed with snow and sleet, the water levels are higher and the current stronger than we expected. Paddling hard to keep warm, we struggled to make our way toward the Town of Fort Frances. We were approximately two days away from Fort Frances, and we were looking forward to finishing up the second year of our cross-Canada canoe trip. We had begun this year paddling against the current up the Fraser River and were finishing against the current. On arrival at Fort Frances, we expected to meet up with Carol's brother Paul who would be making a two-day trip by car from Kingston to transport us home.

At midday, we saw an approaching speed boat along the far side of the river. Once the driver spotted us, the boat made a course change and headed straight for us. As we had been travelling for several days along the international border, Glenn said, "I wonder if it is an American or a Canadian Customs officer checking on us." Our thoughts went directly to digging through our gear to get our passports out. We were shocked and surprised when we found out it was not the border patrol but Paul and his friend John. While we were exchanging quick hellos and laughs and telling them how great it was to see them, the current was carrying us backwards. We declined their offer of being towed to Fort Frances. We stroked onwards with our destination in focus. Paddling with renewed determination, we realized we would not have to spend another night in the cold. We had planned to complete the section from Vancouver to Kingston all in one season. However, with one delay after another, we were nonetheless happy to make it as far as Fort Frances for this season.

Year Three: Fort Frances to Kingston, Ontario

June 6, 2019 to September 19, 2019

Boundary Waters
Fort Frances to Thunder Bay, Ontario

Dates: June 6 to June 26

Route: Boundary Waters; Hwy #588; Kaministiquia River

All winter Glenn had been nervous about paddling on the vast ice-cold waters of Lake Superior. There were times when he would sing a verse from the iconic folk song by Gordon Lightfoot, 'The Wreck of the Edmund Fitzgerald'. He would repeat in his best rendition: *"The legend lives on from the Chippewa on down of the big lake they called Gitche Gumee; the lake, it is said, never gives up her dead, when the skies of November turn gloomy."* This was Glenn's weak attempt to inject some humour to alleviate the stress he sensed. And with good reason. The lyrics of the song portrays so much of the mystery and danger found in the great lake. Lake Superior, called Kitchi-gummi, in Ojibwe, meaning 'great lake' or 'great water', is notoriously known for its violent storms and its icy tomb-like waters. The lake's water is cold enough year-round to inhibit bacterial growth. Bodies tend to sink and never resurface. Vowing we would not become part of the collection we swore to each other to respect Mother Superior at all costs.

We checked and double-checked our gear. It was time to go. We had only Ontario left to paddle! With the canoe secured to the roof of the car and accompanied by Paul and Denise, we headed for the two-day drive back to Fort Frances. Over the winter we had made a connection in Sault Ste. Marie to drop off a food cache. On the waterfront where Lake Superior meets the St. Marys River, we stopped briefly at Catherine and Doug's home to drop off our supply. The City of Sault Ste. Marie, near three of the Great Lakes, is exactly halfway home from

Fort Frances and we had planned to pick up the cache once we had finished paddling Lake Superior.

After completing the drive to Fort Frances, we headed to Pither's Point Park on Rainy Lake to launch our canoe. A small gathering of supporters gave us a fond send-off. Along with press coverage from the local newspaper reporter, Paul, Denise and their friends John, Cheryl and son Jake, our gracious hosts in Fort Frances, were all in attendance. To our delight, others who had befriended us last year along our route also had come to bid us farewell. Loving Spoonful wanted to ensure we looked good. They had gifted us with matching blue T-shirts, emblazoned with our logo, a character likeness of us paddling head-on into the wind. Now it was up to us to finish what we had started to do two years ago. Like the depiction of our logo, we launched into Rainy Lake head-on into an east wind.

Excited to be on the last and final leg of our journey, we headed towards the Boundary Waters. We were looking forward to paddling through the protected wilderness straddling the Canada–United States border between the province of Ontario and the state of Minnesota. Paddling the Rainy River last year, we never once had looked at our GPS as our focus was on reaching the end of the river. If we had, we surely would have noticed that our GPS only displayed the topographical maps of Canada, not of the United States. We could have taken many shortcuts this year among the islands located in Minnesota. However, we could not verify our location without the necessary maps. Not wanting to get hopelessly lost, we played it safe and stayed on the international boundary line with a clear view of Canada on our GPS. There were times, however, where a portage or a campsite only existed in Minnesota. In preparation last year, we had called United States Customs and Border Protection Services at the City of Baudette, Minnesota Port of Entry to ask them how we should proceed through the upcoming Boundary Waters. We had explained that we would be crisscrossing the border between Canada and the United States. It would be impossible to contact or visit their Customs office each time as required by law. The supervisor advised us to download the application 'Roam' onto our mobile phone and to register our passport information. Roam eliminates the requirement to visit a Customs and Border Protection Services office physically. In our case, Roam would track our location through the application's GPS. We felt reassured that we had done our due diligence as it would not have boded well for us were we to be caught illegally in the United States.

Our travels took us over two railroad portages at Loon Lake; one near the start of the lake which happened to be in Minnesota and another at the far end. For a fee of ten dollars, the operator will transport a small boat up and over the steep hill using a rail system powered by an old flat-head car engine. This was similar to the self-propelled trolley we had encountered last year on Turtle Lake in Ontario. The Loon Lake trolley was a slightly different design. We paddled up to the loading cradle, removed ourselves and made sure the canoe was secure. Communication is required to contact the trolley operator. He is perched next to the engine, high atop the hill. Following instructions scribbled on the sign, we walked up to an old numeric keypad telephone mounted on a large pole at the end of the dock. We lifted the handle from the cradle and punched in the required digits. The phone was so old that we had to shout into the receiver for the operator to hear us. After we had given the 'all-okay', the operator started up the engine pulling the cradle the length of the rail line up and over the peak of the hill. Once the cradle got over the top of the hill, the brakes engaged as gravity pulled it down to the other side into the lake. The process took about twenty minutes. We gladly paid the operator for the service and continued on our way. The second trolley portage configuration was similar to the first. There was one common denominator. Both trolley operators spoke rather loudly to us. We assumed they were not upset with the effort of providing passage for our canoe but rather because the constant noise of the old engine had taken a toll on their hearing. This was a good thing for them possibly as Glenn was singing 'The Wreck of the Edmund Fitzgerald' again.

We had often talked of Quetico Provincial Park as a paddling destination. Located in northwestern Ontario, Quetico is the province's first wilderness class park and is recognized as one of the world's premier paddling destinations. Nowhere else in Ontario can you visit a remote ranger station on the international border as you paddle from the United States into Canada or vice versa. Before we had retired from our work, we had known the distance to travel back and forth to the park was out of reach for us during our one-week summer vacation. Now it was a reality. We were delighted to make the destination finally a possibility. Before we could register at the Lac la Croix Ranger Station at the entrance of Quetico Provincial Park, a fierce storm approached quickly from behind. Since our immediate concern was to leave the water and find shelter, we headed for the closest landing possible—a beach near the community of Lac la Croix. As it turned out behind the sandy shore, lay the Lac La Croix Anishinaabe Nations Pow Wow site. To our relief, we found a wooden concession booth used for Pow Wows that was unlocked and empty. This allowed us to take shelter from the raging thunderstorm. We used the concession booth to cook our dinner and wait out the storm before we pitched our tent outside. We were grateful once again to our First Nations peoples for sharing their land. Before we had left our home in Kingston, Carol had made a memorable trip to a nearby First Nations community to obtain a pouch of pure tobacco. We intended to gift the tobacco to pay our respects, to honour the spirits, and to ask for guidance and blessings through these waters. The next morning, before we set out, we left our first offering of tobacco and whispered a prayer of thanks to the spirits for the use of their sacred land.

In Quetico and on Lake Superior, we would be paddling on traditional routes still used by Indigenous peoples. Here, where indigenous activity dates back nine thousand years, we were respectful of the customs of the First Nations and their people. Quetico's naturally existing elements of fish, animals, insects and plants have provided for the Anishinaabe Nation and Indigenous peoples before them. These elements give them their medicines, food, water and way of life for sustenance still to this day. The people of Lac La Croix First Nations have opened their hearts and spirits to the visitors of Quetico Provincial

Park. Here visitors can enjoy everything the park offers as their ancestors did before them. A sense of pride overtook us to know what a privilege it was to travel on these same waters as so many had done before us.

The two hundred fifty-kilometre historic route from Lac la Croix east to Lake Superior, which is known as the Boundary Waters Voyageur Waterway, is a Canadian Heritage River System. We were in awe of the landscape. At times the river was only metres wide. Just within reach of our paddle, the shores to two great nations, winding and meandering, revealed new sights around every bend. Here we paddled past rugged coastlines, vast forests, campsites centuries' old, abandoned mines and quiet protected lakes choked off by wild rice. On the majestic towering cliffs one could find indigenous mystic paintings. The Canadian Shield which is the largest mass of Precambrian rock on the face of Earth contains many indigenous, spiritual pictograph sites. These were teaching areas that were used by local shamans. The pictographs depict art from hunting, forays to powerful medicine symbols. One could imagine a shaman standing in his canoe applying in red ochre paint the drawings that have lasted thousands of years. We were careful not to disturb these sites. We paddled by them respectfully, leaving our gift of tobacco in the waters and giving thanks to the spirits.

Anyone who wishes to get a feeling for the unique history and geography of this country can do no better than explore Canadian water voyageur routes. These historic waterways, travelled by our First Nations peoples and later by voyageurs, extend all across Canada and are a great way to see the diversity of this great country. In addition to Canadian water voyageur routes, there are forty-two designated Canadian Heritage Rivers Systems. We were to paddle seven of them upon completion of our journey. To travel these routes, one must understand this country's watershed and where the dividing ridges are between drainage areas. For example, from Lake Winnipeg, Manitoba to forty kilometres west of Thunder Bay, Ontario the water flow heads north towards Hudson Bay. When planning our route, we had known we would be paddling against the current and that the challenge would be notable on narrow rivers and lakes. On narrow rivers, the current is more noticeable and considerable effort is required to move forward.

Then you have the portage work. This region was characterized by a vast network of waterways and bogs within a glacially-carved landscape of Precambrian bedrock covered by boreal forest. The numerous portages being well-worn provide evidence that this has been a transportation highway for thousands of years. It takes three trips (back and forth) for us to complete a portage. Therefore, if a portage is one kilometre long, we travel a total of five kilometres. Usually, on our shorter excursions, we would take one large pack which Carol would haul and one small pack which Glenn would shoulder while also carrying the canoe over his head. Both of us would then march down the portage trail and complete the portage in just a single trip. Because of the length of our cross-Canada journey, we have more gear than usual. On this journey, there was an average of four to five portages per day. Each portage requires us to haul one hundred ten kilograms of equipment in stages over rough terrain which often has huge boulders deposited by glaciers eons ago.

Carol Reflects:

After a week of portaging, I felt as if my joints were disintegrating. As familiar as I am with portaging, no amount of physical training could have prepared me for the exertion of this challenge. I can honestly say it was easier to walk across the Rocky Mountains than to lug our canoe and gear over roots, boulders and crevices. My body ached in places that I never knew existed, a constant reminder that I am not twenty years old anymore. Portages during wilderness tripping are normal and I am familiar with this type of portage. The path is, at times, muddy. Other portages saw us scrambling over steep rocks. Because the canoe is too cumbersome for one person to carry due to its length and the placement of the spraydeck, we both must share in this task. Without a free hand to swat the mosquitoes, we donned our bug jackets and gloves to avoid the onslaught under the canoe. Grabbing one end, Glenn did the same. Balancing the canoe on our heads, off we marched. We laughed at ourselves and each other as we made noises, grunting and groaning under the pressure of its weight, something we noticed we now do more often. My neck and shoulders ached under strain, but it is all part of the adventure. I would not give up this chance-of-a-life-time experience for anything in the world. I can relate to the quote of the late filmmaker, canoeist and conservationist Bill Mason 'Portaging is like hitting yourself on the head with a hammer—it feels so good when you stop.'

As hard as portaging is in this area, one cannot even begin to compare it with the hardships men endured back in the days of the voyageur. Like we do, each man carrying supplies and furs would also make three trips to complete a portage. However, each voyageur's pack weighed roughly forty-one kilograms, and each man was responsible for six of these packs. That is two packs for each trip for a total weight of eighty-two kilograms! A far cry from today's waterproof bags containing lightweight equipment, quick-dry clothing and vacuum-sealed food.

Glenn Reflects:

On one portage I was first, leading the way down the trail which turned into muddy and slick conditions. I shouted back to Carol to be careful with her footing when I slipped and sprained my ankle. The ankle became swollen and quite sore. It left me hobbling for days. Fortunately for us, we were carrying a well-stocked first aid kit. We were equipped for infections, giardiasis, pain, wounds and anything else we hoped we would never have to encounter. The upcoming days seemed filled with portages. One was three kilometres long, which required me to endure walking fifteen kilometres while carrying equipment to complete the portage. I carefully paced myself, grimacing in pain with each step.

Our route took us into a small narrow opening called Arrow Lake, thus eliminating the Pigeon River where the Grand Portage exists. The famous Grand Portage is the most strenuous of the twenty-nine portages from Lac La Croix to Lake Superior. The portage would have required us to cover the fourteen-kilometre distance five times. So our logic was to push our loaded canoe along a secondary road to Kakabeka

153

Falls, north of Thunder Bay, for forty kilometres rather than carry gear a distance of seventy kilometres over rough terrain. Below Kakabeka Falls the Kaministiquia River flows into Thunder Bay, the sunniest city in eastern Canada with nearly two thousand two hundred hours of sunlight each year.

What a challenging week for both of us! Not only did we have the physical challenge of portaging so many kilometres, but it would turn out to be an emotional week as well. Carol's mother notified her that her father was in the hospital. Family is important to both of us, and we were to re-evaluate our journey once we had reached Thunder Bay. Only a few days later, we also received the news that Glenn's Aunt Marion had passed suddenly. Feeling disconnected from our family was difficult for both of us. Marion was a loving aunt to Glenn. Marion, along with her husband John, was a member of our home support team. We would miss this dear lady.

With heavy hearts, we continued walking our canoe down the highway to the Kaministiquia River in Nolalu. Along the way, we were offered accommodation on private property. Patrick and Emma, a couple expecting their first child, were tree planters living in a rustic hand-built home. This incredible, young couple were entrepreneurs who have planted over fifteen million trees not only in Canada, but in Scotland as well as in Brazil. They live off the grid, homesteading while spending their winters in South America. They create art and use their skills as scuba diving instructors. They had built their one-room home from materials donated by neighbours and from repurposed construction materials. We both admired their sustainable way of life. Even though their worldly possessions were meagre, they lived a rich and prosperous life. Many people we have met in our travels are much like Patrick and Emma—those who have the least—give the most. It restores one's faith in the kindness of humanity. After they had given us a tour of their forested land, Emma made us a delicious stew. We then settled in for a cozy night of rest and conversation. They kindly offered their large, tree-planting travel trailer for us to use for the night.

Patrick had offered to drive us to the head of the Kaministiquia River but we politely declined. By noon, we reached the Kaministiquia River and found it to be a fast, shallow, meandering river, demanding tight manoeuvring through rock-garden Class I rapids. We also had a heart-stopping moment through a Class II rapid which dropped through a series of shallow ledges. One wrong move would have inflicted severe damage to our Kevlar canoe—a most exhilarating experience yet also nerve-wracking. The river flows to Thunder Bay but not before passing Old Fort William. Thunder Bay was born in 1970 of the amalgamation of Port Arthur and Fort William. Here we spotted many large northern voyageur canoes tied up at a dock that leads to the fort's entrance.

During its time, Fort William was the centre of the northwestern transshipment point for furs and trade goods. One cannot help but think back about days-gone-by. Men travelled with heavily laden packs. The men worked in unison singing, paddling and thinking about their break when they could light up their pipe. These were the early traders of the day. The voyageur's daily routine consisted of sixteen-to-eighteen-hour days. The men were to be on the water by three in the morning. They stopped at eight o'clock for breakfast. A midday lunch was no more than an opportunity to eat a piece of highly nutritious pemmican (dried buffalo, caribou or moose) to chew along the way. The stops were fairly regular, so each hour the men would have a few minutes to smoke their pipe. The men would establish camp between eight to ten in the evening and then cook supper usually consisting of beans, corn and peas. What a gruelling routine compared to our nine-hour days!

More than likely early voyageurs had camped in the very same spot next to Fort William property where we pitched our tent on an island. Early voyageurs putting in long days must have had a challenge trying to get well-deserved rest. They did not have a modern-day, lightweight, bug-proof tent to sleep in; instead, the early voyageur dropped down on turf, moss, or beach with their head under the overturned canoe. A tarpaulin would be stretched from the canoe to give shelter from rain and dew. Ordinarily, no provision except the voyageurs' grease and dirt was made for protection against insects.

Leaving a piece of history behind at Fort William, we paddled back into the present as we headed into Thunder Bay at the end of Kaministiquia River. What a striking difference met us as we paddled past the industrial section of the city. Alongside the Great Lakes shipping docks we saw hulking freighters used for transporting materials for the farming, forestry and mining industries. We paddled past towering grain elevators and storage silos. They store hundreds of thousands of tons of wheat, canola, soybeans, flax and oats, awaiting transport by ship or rail.

In less than three weeks we had made it to Thunder Bay. Glenn's ankle was better, no longer many colours of the rainbow. After this year's initial two weeks of building up our endurance, we were once again in great shape. We had planned to spend a day to rest and resupply in the city before we were to tackle Lake Superior.

Carol Reflects:

Unfortunately, I had a sad turn of events. In the city, I received a call from my brother Paul saying that my father was in the hospital and that his health was deteriorating rapidly. A brain tumour had advanced more quickly than expected. The prognosis was not good. Glenn and I decided to fly home from Thunder Bay so I could spend time with my dad during what, I did not know at the time, would be his last days. All his life up to the age of eighty-eight, my dad had been an active, healthy man. Just the year before we had speed skated together at the local rink and gone for coffee at his favourite café. It was hard to see my father now helpless from the disease that had consumed his mind and body. In my world he was invincible. When he had been diagnosed earlier in the year and was still lucid, I had taken the opportunity to say that I loved him and how much he meant to me. After my father passed away, we celebrated his life. Once Glenn and I had picked up our paddles to continue our journey three weeks later, I was able to find peace on the water. My father appreciated the beauty of nature and the wonders of the great outdoors. I had the opportunity to reflect on the fond memories I had of him.

Now that we had picked up where we left off, we felt eternally grateful for the many people who had helped us during this difficult time. We reflected on family and friends who had supported and helped us—friends who arranged our flights and Glenn's past colleague and his acquaintances for storing our canoe and equipment in Thunder Bay. All the help we received, no matter how big or small, had been welcome especially for Carol as she was overwhelmed and emotionally drained during those past several weeks.

Northern Lake Superior
Thunder Bay to Pukaskwa National Park, Ontario

Dates: July 15 to July 23
Route: Lake Superior

Thunder Bay is on the spectacular coast of Lake Superior. Now paddling in the peace of the early morning calm, we cannot explain the feeling we got when seeing Lake Superior for the first time. Lake Superior is just that—superior in every way. Heading our little canoe out onto the lake, we felt small and insignificant compared to the vastness of what is the largest freshwater lake in the world. Lake Superior is considered an inland sea. The world's largest freshwater lake by surface—an area the size of Austria—Lake Superior has a maximum depth of four hundred five metres. It is so large it creates its own weather patterns. It is well known for violent storms with ten-metre waves that can sink colossal steel freighters, not to mention tiny canoes. The cold climate year-round is so harsh that some species of plants commonly found in the Arctic climate grow here. Looking out at Lake Superior, we could only anticipate what great adventures would await us!

We were in awe as we paddled towards Sleeping Giant. The Sleeping Giant is a formation of flat-top mesas. It juts out into Lake Superior and resembles a giant lying on his back with arms folded across his chest. It is a sight to behold indeed. At two hundred fifty metres in height, its dramatic cliffs are among the highest in Ontario. There are many stories about this landmark. One Ojibway legend identifies the giant as Nanabijou, the benevolent god who watched over the tribe and turned to stone when the secret location of a rich silver mine, now known as Silver Inlet, was revealed. We took the opportunity to camp

159

in a sheltered cove at the base of the giant. Ours was one of the many backcountry campsites scattered within the hiking trail system. Even though we could not see the giant this close up, we did feel his presence in this mystical place.

Calm weather prevailed as we paddled east of Sleeping Giant to Edward Island along Magnet Point passage to Shaganash Island lighthouse. At the lighthouse, there is a commemorative sign stating we are travelling on a section of Canada's Great Trail. The Great Trail is also part of an ancient heritage highway: Path of the Paddle. This section is nearly one thousand kilometres in length and goes from Thunder Bay to Sault Ste. Marie. Much like the moody goddess of Lake Winnipeg, Lake Superior is said to resemble a wild beast. The lake can be sunny, bright and beautiful one minute and all rolling clouds, gusts of wind and torrential rain the next.

Just as we were paddling away from the lighthouse, the beast became unpredictable in another kind of way, fog. We could see great banks of fog rolling in. We became entirely engulfed as we crossed Swede Island channel. We tuned into our marine radio to check the area only to find, to our dismay, the device was broken. Using the GPS to guide us and with no visual sight to focus on, Glenn would call out to Carol to move either to the left or right. Without having had the GPS to guide us, we would have easily paddled in circles. As Glenn stroked, he concentrated on the heading. He directed the canoe through several massive fog banks until after a couple of hours the sky finally cleared. Fog of this type was a phenomenon we had never before encountered. Not being able to see more than a few mere metres in front of us gave us an eerie feeling. In the stillness of the morning calm, it was eerier still to hear the water lapping on the shores of the small islands, yet we were unable to see them.

Besides storing our food cache, Catherine and Doug from Sault Ste. Marie had given us an invaluable gift. Having themselves paddled the north shore of Lake Superior for many years, they correlated a detailed map with campsites and points of interest along our route. There was not a day that went by where we did not appreciate this gift of kindness. One of the tips they had given us was the location of a rustic sauna camp located between Loon Lake and Borden Island. This sauna is one of many that are popular along the northern coastline of Lake Superior. To heat the rocks, each sauna has a stove in which to make a fire. Then throwing water on the rock creates steam. After heating up in the sauna, naked as a couple of jaybirds, we jumped into the ice-cold waters of Lake Superior. Such an invigorating feeling after a day of paddling.

Lake Superior is cold and unpredictable. Although water near the shore may be warmer, the lake's year-round average temperature is only four degrees Celsius. Hypothermia can set in within five to ten minutes if one is not wearing a wetsuit or drysuit. We had neither. After our first year of this odyssey, we had adapted from a traditional personal flotation device to a more comfortable inflatable type. The lightweight vest is easy to wear and has a manual inflation handle. If the situation arises that we need to deploy the vest, we must pull the handle to inflate. Best practices dictate to stay with a boat once it capsizes; in this cold lake, however, staying in the water with our canoe, we would not stand a chance of survival. We had already discussed that, in the event we did capsize again, we would swim quickly to shore. Regardless, we were not taking any risks of overturning in this lake. Once the waves started capping, it was a sign to get off the water. When cold water meets with warm air, weather changes swiftly and two-metre swells can come up in short order. We had to be aware of rip currents—hazardous currents caused when river currents mix into a lake current and produce unpredictable patterns. Using the GPS, we kept a watchful eye and gave all rivers wide-berth all along this coastline.

We had a day that continues to replay over and over in our minds. Conditions appeared fine; we were both relaxed, having easily crossed the distance of Shesheeb Bay. We even stopped for a lunch rest and a quick cold dip in a bay at Spar Island. The winds picked up slightly as we headed out through the strait that empties from Nipigon Bay. Crossing at the mouth of the Nipigon Straight, we could see small choppy waves from a distance. Now in the middle of the channel, to our dismay, the small choppy waves became more massive, rolling waves. We had misjudged and forgotten the dangers of channel cross-ings on Lake Superior. We found we were now in the middle of what we had both feared. Adding to our fear, we could hear the waves crash-ing along the cliffs of the upcoming shoreline of Fluor Island.

With no chance to turn back, we both agreed to set our sights on the leeward shore of a small cluster of islands on the other side of the channel. Our nervousness grew by the minute as the wind suddenly increased. We were adjusting the direction of our course slightly to paddle with the waves as the wind buffeted our canoe. The water became a boiling cauldron. Struggling to compensate for the wind which was driving us toward the rugged coast, Carol maintained order with great effort. From the corner of our eye, we noticed a sailboat holding back, keeping watch, we assumed. Riding the crest of each wave, fighting the urge to panic and keeping a clear head while pushing thoughts away of falling into the frigid water, we continued to stroke steadily. After an exhausting twenty minutes which seemed like hours, to our great relief, we made it intact out of the channel. Having reached calmer waters, the sailboat, which was also struggling to remain in sight of us, quickly vanished. We paddled to the leeward side of the small islands we had set our sights on and rested for a few hours in the safety of the protected bay. We were both in mild shock as we reflected on how close we had come to capsizing. We swore that we would be more aware of upcoming pitfalls. Discussion ensued on the dangers of becoming over-confident and how easy it is for mere humans to think they are invincible.

The islands were temporary protection from the wind but not suit-able to camp on, especially if the winds were to pick up. We built up our courage and set out once again, leaving the turbulence of the channel behind us. Now into our fifth day of paddling on Lake Superior, we had seen less than a handful of people. So, when we stroked past a cove and saw the mast of a sailboat, we assumed right away it was the same vessel we had seen in the dreaded channel. Paddling up to the couple in the sailboat, we had it confirmed they had indeed had held back to keep a watchful eye on us. The sailors said they had had struggles of their own as the aluminium fishing boat they were towing kept wanting to overtake and pass the sailboat. But they said they felt a need to hold back in case a rescue was in order. The skipper then mentioned, "Obviously, it was not your first rodeo with that canoe." It is reassuring to know that marine travellers watch out for one another and realize that in a moment's notice, the situation could change for the worst. This lake is not a place for inexperienced boaters. The lake has taken many lives, even those with experience.

Crossing over to Battle Island, we wanted to explore the lighthouse positioned high up on a cliff. Many stories of Battle Island surface. A November storm which raged for three days in 1977 produced one hundred thirty kilometres per hour winds. Waves built to a height of fifteen metres as the waves travelled from Duluth, Minnesota. When they hit the rocks below the lighthouse, they were strong enough and high enough to wash up the tower which is thirty-seven metres high. According to former lighthouse keeper Bert Saasto, "They smashed the glass right out of the lantern. That is what puts the fear of God in you: the sound of the wind and the waves. You cannot get away from it. You almost panic, and you feel like running but, out here, there is nowhere to run to."

We entered a cove on Battle Island and met a group of eight kayakers boisterously singing 'Old Man River'. We soon discovered the group had met during a kayaking excursion a few years ago. They had formed a common bond. They enjoyed each other's company, and every year they came together to paddle different sections of the coastline. The group was finishing up their annual trip and heading to the mainland at Rossport. We chatted with ottO (yes, that is how he spells his name), the group's leader, and with the others in the group. On the water, a common bond is often shared by fellow paddlers. We exchanged stories of water travel; adventures experienced and tips of any new and fancy camping gear now on the market. Upon hearing that our marine radio had broken, ottO, without hesitation, unclipped his costly VHF marine radio and handed it to Glenn. He explained that we were going to need it more and did not seem concerned about its return. We insisted that we would give it back to him when we passed his hometown of North Bay which was on our route. Another member of the group gave Carol a three-litre bag of leftover wine. She was ecstatic with this gift. We were aware of one-litre bagged wine before but a three-litre? It tucked nicely behind Carol's seat and, from that day forward on the trip, she always travelled with a bag of wine. The wine had suddenly become another, added luxury item on any future canoe trip.

'Luxury items' on a canoe trip can mean different things to different paddlers. Luxury items are not a necessity, like rain gear or safety equipment, but can make a real difference to the enjoyment during a trip. When we go paddling for one week, one month or three seasons, our luxury items are based on the type of journey we take. If enduring portaging, we take very few luxury items; however, on water routes where portaging is not required, like Lake Superior, we load right up. On top of our list is a little Irish Cream to put into our coffee at the end of the day and two lightweight camp chairs to rest our backs. Then add a good book or two which we read to each other or on our own. A recent purchase that one may call a luxury item or a necessity explicitly added for the blood-thirsty mosquito-infested forests in Ontario was a screened-in-shelter. We replaced our tarp for the shelter. Weighing approximately two and a half kilograms, it is a waterproof tarp with no-see-um mesh walls. It is large and can fit eight people inside if need be. There are two doors for easy access, and we can use the tie-back loops to hold back the bug mesh in case we want to use it just for shade. The shelter is heavy to slog over portages, especially considering the weight of the other luxury items; however, we would not set foot in the woods during the spring without this shelter. It gives us a place to escape the bugs while we are sitting in our camp chairs, reading, enjoying our laced coffee or, now, wine, at the end of a long day.

After we had paddled in the oscillating wave action of northern Lake Superior, the effects of these pitching waves still lingered after we stepped onto land. Both of us were off-balance, dizzy. Bobbing up and down in the water for so long made us feel, even on firm ground, like we were still rising and falling with the waves. This sensation happens every time we paddle in swells or rough water and the effects on us linger well into the night. Because the conditions of the water range from calm to wavy and are different every day, one does not get used to this off-balance feeling. At least we did not. It seemed every time it happened, we had to get used to the sensation all over again.

A gentleman approached as we were setting up camp at the mouth of the Steel River. He referred to himself as 'Steel River Jim'. Originally from the east coast, he had now found a place on Lake Superior to call home. We passed the evening by enjoying Steel River Jim's unique and genuine character. We were eager to learn from this fellow paddler and kindred spirit. Steel River Jim told us how he needed to find himself one summer and set out travelling solo by canoe into Nipigon Bay. Chuckling, he told us that he had equipped himself with a sail rigged from a shower curtain to assist him when the wind was at his back. When there was no place to pitch a tent on cobblestone beaches so typical on Lake Superior, Steel River Jim would secure his canoe on top of two logs placed side by side on the beach. He firmly wedged the canoe to prevent movement. With only a canvas tarp laid over the top to protect him from the elements, Steel River Jim would commence sleeping in the bottom of his canoe. He also told us this trick was an excellent method to launch a fully-loaded canoe into waves. Later we tried this method, not to sleep in the canoe but to launch it from the shore using two logs. Much like Steel River Jim suggested, near the water's edge we would rest the canoe on top of two well-placed, driftwood logs. We would load the gear into the boat and then slide the canoe into the water and jump in. This method was so much easier than trying to load our equipment into a canoe bouncing in the waves.

The coastline is ancient. There is something so revitalizing about being in such a vast open space surrounded by nature. The crystal-clear waters, the terraced stone, sandy beaches and cliffs are all part of the Superior experience. Before crossing the expansive opening of McKellar Harbour, Carol spotted, high up behind a terraced stone outcrop, what looked like pools of reflective water. She climbed out of the canoe to discover smooth rock basins filled with water and warmed by the sun. An invitation for us to take a dip. Not deep enough to swim in but refreshing enough to bathe in; easily ten degrees warmer than the lake. Drinking in the view of the vastness of the calm lake before us, we explored the area.

Our destination was the narrows between Foster Island and the mainland just past Pic Island. The sweeping topography of Pic Island has inspired some of the most famous paintings by Canada's Group of Seven artists. And, no wonder, we thought as we paddled through its stunning, hidden passages of the inner islands, home to some of the oldest rocks on earth. The smooth rock shores of Foster Island enabled us to pitch our tent in a beautiful cove directly on the flat rock. We had supper and relaxed by a small fire constructed in a crevice of a rock ledge. After a pleasant evening, we went into the tent as the night had grown cold. The bedrock of the island beneath us was still warm from the strength of the sun rays during the day. Life is good.

Northeastern Lake Superior
Pukaskwa National Park to
Michipicoten Bay, Ontario

Dates: July 24 to August 2

Route: Lake Superior

Down the coast from the Town of Marathon is Hattie Cove and the west entrance of Pukaskwa National Park. We took the opportunity to register with the park and acquire a map of the coast. Pukaskwa National Park from Hattie Cove to Michipicoten Harbour is a two hundred twenty-five-kilometre wilderness paddling route. The coast comprises the longest undeveloped shoreline in all of the Great Lakes. Here we were to discover incredible geology: powdery sand and colourful cobblestone beaches, beautiful wilderness campsites, breathtaking waterfalls and sacred Indigenous sites. Numerous times we have been told the park has an unforgiving coast with few places along the shore that would harbour a boat during strong wind. However, the benefits of paddling this remote shoreline outweighed the negatives. With the recent memory of our experience at Nipigon Strait, we were wary of crossing the opening near the mouth of the White River. Naturally, we assumed it had strong currents and undertows, so we kept a safe distance from the river opening.

While at the Hattie Cove Campground, we happened to strike up a conversation with a couple from Mattawa, Ontario who, over several years, had paddled most of the northern shore of Lake Superior. Lynn and Mike gave us their information and insisted that we contact them once we had arrived in Mattawa later in the summer.

Glenn Reflects:

As we were preparing to set out the next morning, Carol stumbled from a small ledge leading to the water and sprained her ankle. I am concerned she may need medical attention; she is pale and in discomfort. Carol insists on continuing anyway.

Carol Reflects:

Finishing this journey without any further delays is at the forefront of my mind. I took off my boots and hobbled immediately into the numbing, ice-cold water of the lake. My right foot became swollen and days later turned black and blue. I was not able to walk without the aid of a paddle or a stick. As had happened when we were in the east coast of Canada, I felt badly for Glenn once again burdened with more camp duties. With my recent diagnosis of Osteoporosis, a common outcome of Celiac Disease, I wondered if I had broken my ankle. Sending healing thoughts towards my foot and trying to stay positive, I was thankful most of the travel along this coast would be in the canoe. To encourage healing and to ease the throbbing sensation, I elevated my foot. When the opportunity arose, I would also drop my foot over the side of the canoe in hopes the cold water would ease the swelling.

Today was July twenty-fifth. For the most part, the winds had been relatively low since we had left Thunder Bay ten days ago. This gave us peaceful waters on which to paddle. Each day we arose between four and five in the morning to get on the water a couple of hours later. Usually, by two or three o'clock in the afternoon, the winds would pick up, and we would be looking for a campsite. The marine weather radio had reported a strong wind warning for the following days. We found a sheltered cove at Morrison Harbour and waited out the wind. Shortly after we had erected camp, a sailboat appeared in the cove and moored in the deep waters. A dingy dropped from the stern of the boat and a lone occupant rowed toward our beach. Jean explained that she was on a sailing holiday for a couple of weeks with her father. They had listened to the same forecast and decided to hide behind the islands for a day or two until conditions had calmed down. We shared a cup of coffee while we told tales of our adventures on the water. Like we had from Steel River Jim, we learned another little trick from Jean. She sprouts lentils by putting them in a jar with a bit of water. She then places the jar in a warm, dark place and rinses them twice per day. After four days, the lentils produce sprouts—green crunchiness to add to meals. Carol later experimented with lentils in a resealable zipper storage bag and placed them in a warm spot under the spraydeck. Days later we enjoyed sprouts!

The anchored sailboat in the bay was a good indication of how stiff the wind was. Pivoting with the direction of the wind, pulling tight like a fishing line in the current, the vessel acted as a weather vane. The winds continued throughout the evening and well into the next day. Thankful to be safely on land in this well-protected little cove we peered out of the tent. The sailboat was tossing and rocking about with the motion of the waves. We made the best out of the next day by continually listening to the marine radio. Every hour we hoped magically for a turn in the forecast. At the same time, we explored a portion of Pukaskwa National Park's shoreline hiking trails behind our campsite. Carol gathered smooth rock tripe (*Umbilicaria*) a lichen which is found on boulders and rock walls. She intended to add it to our stir fry. After soaking the smooth rock tripe in baking soda water for several hours, she then fried it with reconstituted dehydrated vegetables and spices and served it over rice. Lichen is a good source of Vitamin C which had been lacking in our diet for some time.

To pass the time we relaxed in the sun in our lawn chairs on a sandy white beach, we read and caught up writing notes in our journal. The rest was beneficial for Carol as it allowed her to recline and elevate her foot which was now looking like the many different shades of an artist's palette. Following the third night of being stranded on the beach, we were ready and eager to continue our journey. We powered on ottO's marine radio only to find we had inadvertently left it on all night, thereby draining the battery. The radio required a charging system which, unfortunately, we could not mate with our solar panel. When we had finished breakfast, we heard the sailboat's small engine firing up and looked at each other, wide-eyed. We jumped into the canoe to paddle out to the vessel and inquired about the marine forecast. Upon hearing that our marine radio was dead, Jean's father offered us his spare to use. The skipper informed us he would send an email with an address where we could mail it back once we had completed our journey. This act reaffirmed once again that boaters travelling on Lake Superior selflessly look out for one another. The weather was to be somewhat stable for the foreseeable future so we made short work of packing and headed out.

Unbeknownst to us and not much further down the coast was a paddler also stranded for several nights. Since this shoreline is remote, we would go out of our way to wave or say hello to a fellow paddler. Before this adventure, to seek solitude, we would paddle out of our way to avoid people. Gregg, a solo kayaker, was travelling south for a day or two before turning back. After discussing and determining we were not invading each other's solitude, the three of us paddled together. Gregg led the way in his sleek sea kayak followed by us in our bulky canoe. Along the way, we were comparing the advantages and disadvantages of kayaking versus canoeing. It seems sea kayaking is the preferred method of transportation on these waters as we met only one other couple in a canoe similar to our outfit. Both have their advantages and disadvantages. For our purposes, on our voyage, the long expedition canoe had more room for food storage and gear. We had the freedom to travel longer and to reach the more remote parts of our route.

Fog is common on Lake Superior and can last for days. We saw it creep in from across the lake. The fog shrouded the landscape in a thick, white veil. Our perception of the coastline diminished as the fog's ghost-like tendrils curled around our boats and surrounded us in dampness. The coastline was rocky; veering too close would put us in danger of being tossed against the jagged granite cliffs. With guidance from our GPS, we were able to travel a safe distance from shore.

We were now guiding Gregg who only had a map. Thankfully, the fog did not last for days. Eventually, we paddled to Fisherman's Cove for the evening where we shared a campsite with Gregg. Fisherman's Cove reminds one of a scene on a picturesque Caribbean Island. The watercolours vary from Caribbean turquoise to deep ocean blue with fine white sand. The cove is surrounded by sweeping cliffs topped with windblown jack pines. A scene straight from paradise.

The following morning, while sharing a cup of coffee with Gregg, we discussed the coastline ahead. Our newly-acquired Pukaskwa map revealed a possible danger from reflection waves at 'The Ramparts' just past English Fishery Harbour. The map stated: "Danger from reflection waves. Wait until calm." Reflection waves form when waves meet a sheer rock wall and bounce back. The waves have no chance to break over the surface; this makes for unpredictable conditions. If the winds were unsettled before we reached 'The Ramparts', we would have to find a location to wait it out. If all remained favourable, we would pass by this spot on this day with our sights set on a designated campsite at Cascade Falls. Gregg was undecided at that point as to whether or not to attempt this section of the coastline. He talked about the possibility of heading back. We said our goodbyes and eagerly took to the water in hopes of putting in a full eight hours or more. We silently wondered what the lake would offer today.

The kilometres melted away and we finally reached the 'The Ramparts'. There was a mild wind blowing, and the chop was minimal with the waves no higher than the height of our canoe. We would have preferred dead, calm waters, but we had decided to proceed anyway. With reasonably mild conditions, we found this area a scary place. As waves bounced off the cliffs, they would reflect out, rocking our boat unpredictably. The trick was to stay far enough away from the sheer rock wall to avoid the brunt of the reflection waves yet not so far out into the lake to get caught in rolling swells. There was no place to harbour a canoe in case things went wrong. That is why there was a warning on the map, and that is why we tightened our grip, dug our blades in deep and forged ahead to exit as quickly as possible.

Eventually, after what seemed like an eternity, the tip of Otter Island and its historic lighthouse came into view. Assisting mariners during navigation, the structure is still in operation. Directly across from the point in a sheltered bay is Cascade Falls. We were amazed by the scene before us as we quietly entered the cove. This beautiful three-fingered waterfall, immortalized by Bill Mason, is a must-see for coastal paddlers. The sand and cobblestone campsite at the base of the falls is paradise, according to Mason. After beaching the canoe, we set about exploring the unique area at the bottom of the falls. While the falls have remained unchanged throughout the ages, the beach can change character annually. Winter storms frequently pummel the coast. They reshape the cobblestone and cast sizable amounts of driftwood far from the shoreline. Bouncing off the opposite cliff within the cove, the thundering sound of the waterfalls appears louder than they are. The cobblestone beach was rugged for Carol to walk on with her sore ankle. Carol placed her steps carefully as she negotiated the tumbled mess that had piled up on the beach.

We scouted the area in search of a spot to erect our tent. We chose a small, sandy, open patch between the driftwood debris near an outcrop of rock. Behind our tent was a small treeline hiding a nice patch of grass where we put our bug shelter. Complete with a priceless view of Lake Superior, it turned out we had chosen well. The area behind the treeline served two purposes: giving us protection from the wind and providing a reduction of sound from the falls. A fine mist also settled around the area from the falls, but we were far enough away to stop from getting wet.

The following day the winds increased, and it was here that we spent the next two days. Windbound, yet enjoying ourselves, we were resolved to the fact that we were safe and we set about performing camp tasks. Clothes were washed and laid out to dry on the rocks in the sun. We also decided it was time for a well-deserved wash for ourselves. Superior is breathtakingly frigid but inland rivers running into the lake heated by the sun make for a comfortable backcountry shower. Having not adequately bathed since we had left Thunder Bay, we eagerly stepped under the curtain of the rushing water and washed away days of sweat and grime. Glenn spent time clearing the driftwood off the beach to expose areas where others might wish to place their tents in the future. Carol collected small pieces of driftwood sticks that have twisted and curved shapes. Arranging them to form words against a smooth rock background, she formulated greetings to family and friends and took photos of her handiwork. Pocket novels come in handy as we lounged in our camp chairs and sipped wine in the warmth of the sun. The following day we hiked through the dense bush to view the falls from the top. Along the way, we located an old, wooden trapper's hut missing roof and doors. The remains were covered in thick, rich moss. Stranded on this alluring beach we witnessed a spectacular sunset and again were amazed by the beauty of this area. While the sun was setting and with darkness approaching, we retired early, serenaded to sleep by the sound of the waterfall.

Well before sunrise, we listened to our newly-acquired marine radio. After hearing the day was going to be calm, we quickly set about to break camp. Eager to make some distance, we launched into the early morning water using the 'Steel River Jim log technique'. Torpedoing our loaded canoe off the cobblestone beach into the deep water at the shoreline, we anticipated a great day. We were looking forward to enjoying the fantastic scenery that only this big lake could offer.

Crossing behind Otter Island at the narrowest point, put us in alignment to paddle around a long peninsula at Deep Harbour. The exceptionally calm water allowed us to see cliffs along the shoreline up close and personal without dreaded reflection waves. We had a beautiful feeling of isolation as we neared the southern boundary of Pukaskwa National Park. We knew that we were ninety kilometres from the nearest road. From Pukaskwa National Park into Lake Superior Highlands, we experienced more wilderness, waterfalls, pristine beaches and the stunning rugged coastline of Crown (public) land on our paddle to the Township of Wawa on Michipicoten Bay.

Eastern Lake Superior
Michipicoten Bay to Sault Ste. Marie, Ontario

Dates: August 3 to August 11

Route: Lake Superior

The adventure outfitter in Wawa is a busy hub of activity. They are the only outfitters on this section of Lake Superior offering services such as guiding excursions or paddling instructions. They are a fantastic source of information regarding the coastline. The proprietor asked us if, in our travels, we had been offering tobacco to the spirits. We informed him that indeed we had been. He brought out several maps of the area and gave us locations where an offering of thanksgiving would be well received. Spending time on Lake Superior had made us appreciate its unique energy. Ancient Ojibwe people had painted red ochre pictographs on rock faces in recognition of the lake's power. All the sites are of significant historical origin. We quickly documented the location of his recommended sites as we intended to observe the ancient custom of placing an offering to the spirits.

After we had spent the night at the outfitters' small campground, we paddled towards Lake Superior Provincial Park just past Michipicoten Bay. At the northern tip of Lake Superior Provincial Park was Old Woman Bay. As we crossed the bay from the north, we could see deep into the eastern corner of the cove a long sandy beach populated with driftwood. The face of the 'old woman' watches over the bay. If one stands on this beach looking towards the horizon, within the two hundred-metre standing cliffs to the left, the Old Woman is revealed.

Before paddling to our next location around Grindstone Point, we needed to rest and have lunch. The coastline was steep with few places

175

to access, so when we located, tucked in a cove, a cobblestone beach with an elevated shoreline, we headed straight for it. We found the cobblestone beach occupied by a lone kayaker enjoying the sun and sheltered from the wind. Not wanting to infringe on his privacy but needing to take a break, we called out to him and asked if he would mind some company, at least till we had consumed our lunch. He assisted us up the elevated shoreline to secure our canoe. While we were having lunch, the kayaker sat quietly and listened to the tale of our cross-Canada journey. The kayaker said he had been working at Lake Superior Provincial Park for the past thirty-five years and was taking a few days to enjoy some paddling. Carol asked him where we could register when we exited the park. The kayaker humbly said as the Superintendent of the park he would waive the park fees for all our time spent there. We truly hoped we had not infringed on his solitude when he was probably trying to escape the onslaught of tourists.

Leaving the kayaker, with a warning in our thoughts of reflection waves at Grindstone Point, we surmised the wind was not going to allow easy passage. Just before Grindstone Point, leaving the protection of the leeward shore, we rounded the corner to find a southeast wind blowing us broadside. This wind forced the canoe toward the coast. Once again, having no opportunity to hide in a cove or land safely, we decided it was safe enough to forge ahead rather than turn around and go back. We compensated by angling the canoe into the wind and held our breath as the bow crested with every wave. This course took us further out into the lake than we would have preferred. Here was more significant water, more giant waves and bigger wind taking us away from the rugged rocky shoreline. Although scary, it was still manageable. We intended to mimic a sailing manoeuvre to tack into the wind. Once we had reached the point where the winds would align us to shore again, we pivoted the canoe around. We rode with the waves towards a park campsite past Cap Chaillon, a cape, for approximately two kilometres. Once we had reached land, we both breathed a sigh of relief. We thanked the spirits of Kitchi-gummi with a gift of tobacco that night for keeping us safe.

Was it by chance that the gift of tobacco brought incredibly calm waters the next day? It could not have come at a better time for us to explore the picturesque Gargantua Islands. You do not have to be an expert to appreciate the vast range of geological formations along this coastline. From our canoe, we could see where water had left pit-like depressions. These were caused by swirling boulders grinding into the red sandstone during the receding ice age. A different type of red stone, known as rhyolite, comprises a Mars-like section of barren, scalloped shoreline south of Gargantua Bay. Within this wild and ancient place, we came across two islands called Devil's Chair and Devil's Warehouse. Here we found cliffs and pothole caves overhung with old cedars. Years ago, both islands were named after missionaries eager to abolish the Anishinaabe's most spiritual connections. The stories continue and are being revived and retold by the people whose cultural heritage is rooted here. It was quite fitting to once again grant an offering of tobacco to respect the spirits of our First Nations peoples. Just below the surface of the water, the bay also holds several shipwrecks that one can see while paddling over them.

The shoreline of Lake Superior Provincial Park continued to consist of cobblestone beaches. Between the cliffs, the beaches consisted of large boulders to rocks the size of melons; from pea-sized pebbles to brown sugar-coloured sand. We paddled by polished granite slabs and greenstone headlands surrounding Brule Harbour. They are among the oldest rocks on earth.

Behind the coast, there is easy access to the Voyageur Coastal Trail, which is a hiking path. Here we stopped to have lunch, stretch our cramped legs and had great luck finding fresh blueberries for the next day's breakfast. Paddling along the tugboat channel was rewarding as we took in the spectacular sights of rocks that appeared to be vaulting out of the water. This area of the coast was so alluring that we were disappointed when it came to an end, but thankful to have witnessed it.

The upcoming day's weather was hot, and the winds relatively calm. We saw dark threatening clouds forming and detected the sound of rumbling thunder in the distance. Arriving at Pancake Bay Provincial

Park, we checked our cell phone and retrieved a voicemail message from Catherine and Doug in Sault Ste. Marie. They were calling to inquire if we had a safe haven as violent thunderstorms were in the surrounding area. We returned the call immediately. We informed them we were safe and secure and looking forward to our arrival in the next few days.

Carol Reflects:

Because of the impending storm, we had no choice but to camp at Pancake Bay Provincial Park. I called the park office to inquire if they had, close to shore, a designated tent camping area for paddlers. Like most provincial parks we were told, no they did not. While Glenn watched the canoe, I had already endeavoured to walk to the camp office, but the pain in my ankle forced me back. The camp warden now dispatched to locate us arrived wearing a bulletproof vest. She had a staff member with her for backup. I did not know that, in Ontario, wardens wear bulletproof vests and carry batons. It gave me a sense of foreboding. I wondered if Glenn and I would be safe in an area where staff are militarized to keep the peace. Well, at least now we knew that if our neighbours were to fly into a fit of rage after they had burned their s' mores over the campfire while playing their music too loudly, the neighbours would be reprimanded with force. We were assigned one of the last sites available in the park. We set up our tent on gravel that was meant for a large recreational vehicle.

After a night of rain and thunderstorms, we left as early as possible, shrouded in fog and mist, to head along the coast of Pancake Bay. Our initial thought was to travel out in open water for seven kilometres and pass by the shores of Batchawana Island. The island, located between the two points of land, creates the opening to Batchawana Bay. After we had paddled a short way into the bay, the fog began to dissipate, slowly revealing the outline of the island. Still hoping to make the jump to the other point we paddled toward the island but the further we proceeded away from the shore, the more substantial the winds became. The force of the wind coming off Lake Superior was driving us more deeply into the bay. To compensate we veered off course. The decision was made to round the island at the bottom and, hopefully, come up on the lee side shielded from the wind then cross over the distant point. However, the winds did not cease. This forced us to find shelter. As luck would have it, we spotted a white sandy beach that stretched for about two kilometres along the shoreline.

If one were to be windbound, this was the perfect place to be. It was like being stranded on a desert island. We assumed that, like other islands along northern Lake Superior, Batchawana Island was First Nations or a provincial park or Crown land. But we were surprised to discover later that Batchawana Island is one of the largest private islands in Canada. Completely forested and undeveloped, it appears today mainly as it would have at any point in the last thousand years. This quiet island has witnessed the passing of countless generations of aboriginal peoples, early explorers, fur traders and travellers like we were. All of us had encountered its remote and deserted shores.

On the following day, we were sitting in our camp chairs watching the world wake up while we enjoyed an early morning coffee. A moose sauntered from the forest and walked up and down the shoreline while he munched on water plants. We watched silently and felt a sense of thankfulness for this magnificent moment.

As the day progressed, we could see substantial rolling waves in the channel we needed to cross to continue to Sault Ste. Marie. Fear rose in our bellies. Knowing firsthand the dangers of this raging beast of a lake, we decided to stay put. It was not a wonder that Glenn started humming Gorden Lightfoot's song, "The Wreck of the Edmund Fitzgerald". Lake Superior has claimed thousands of boats, ships and canoes over the centuries. The Edmund Fitzgerald is simply the most famous and one of the most recent. The American Great Lakes freighter sank in a Lake Superior storm on November 10, 1975. The ship was caught in a severe storm with near hurricane-force winds and waves up to approximately eleven metres high. She went down, and her entire crew of twenty-nine men was lost. The ship still lies at the bottom of the lake, and the bodies have never been recovered. Glenn sings, *"Does anyone know where the love of God goes when the waves turn the minutes to hours? The searchers all say they'd have made Whitefish Bay if they'd put fifteen more miles behind her. They might have split up, or they might have capsized, they may have broken deep and took water. And all that remains is the faces and the names of the wives and the sons and the daughters."* The 'Edmund Fitzgerald Lookout' provides an expansive view of this area including Whitefish Bay where the SS Edmund Fitzgerald lies eighteen kilometres due west of Coppermine Point.

As the wind became calm in the night, we decided to leave in the wee hours of the morning. Our alarm woke us at two-thirty in the morning. We had breakfast, packed up and launched our canoe in complete darkness, all before four o'clock. It was August 10. The sky was clear, with no ambient light sources to distract our vision. As with many nights on Lake Superior, the stars were brilliant. The moon had set hours ago, and we paddled in complete darkness with only the GPS to guide us. Once Glenn set a course, Carol focused her sights on one of the stars in the constellation to guide us. We only had to refer to the GPS several times to ensure we were still on the right track as we could not see the shoreline. As we were paddling, a multitude of shooting stars crisscrossed the sky; the heavens illuminated. We found out later that we were paddling in perfect alignment to see the meteor showers of the 'Perseids'. The Perseids, which peak during mid-August, are considered the best meteor shower of the year. With swift and bright meteors, Perseids frequently leave long "wakes" of light and colour behind them as they streak through the earth's atmosphere. The Perseids are one of the most plentiful showers with fifty to one hundred meteors visible per hour. Never had we witnessed such a show of shooting stars. As we stroked, we gazed upward in awe, enjoying the show until dawn.

As we paddled, we watched the sunrise and the world wake up. The winds never materialized this day and, with determination to paddle away from Batchawana Island, we arrived at Doug and Catherine's by mid-afternoon. A steak dinner and a rejuvenating shower awaited us. Two kind people bent over backwards to make our stay most enjoyable. Not only did they wine and dine us, but we had the pleasure of enjoying their outdoor sauna. We again experienced true, genuine kindness typical of so many Canadians we had the honour of meeting in our travels. With Lake Superior behind us and our food barrels replenished with our cache of dehydrated food, it was time to say goodbye to our friends. But before we headed out, Catherine gave us a piece of smooth local slate rock which Denise had admired when we had stopped in Ste. Sault Marie on our way to Fort Frances. Catherine asked if we would deliver the rock to Denise upon our return to Kingston, Ontario. Carol safely stowed it between layers of packed clothing. We were transporting a little piece of the Lake Superior landscape home with us.

Lake Huron
Sault Ste. Marie to Killarney, Ontario

Dates: August 12 to August 20

Route: St. Marys River; North Channel
and Georgian Bay of Lake Huron

With the great Lake Superior now behind us, we paddled through the St. Marys River and the historic Sault Ste. Marie Canal towards Lake Huron. Lying in the northern reaches of Lake Huron is a remote, wild passage known as the North Channel. To the north lies the Ontario mainland; to the south is Manitoulin Island, the largest freshwater island in the world. The days pass, and we see an increase in traffic of large, expensive yachts and sailboats. It is evident to us why sailors and boaters find the North Channel one of the best freshwater cruising grounds in the world. One can find harbour in the evenings between the smooth, rocky shoreline of hundreds of uninhabited islands, countless coves and hidden beaches protected from the expanse of Lake Huron. Seeking shelter ourselves in this pretty area of the world, we continued onward and located a smooth and tapered rock ledge on an island. Just large enough to erect a tent, the location included a spot right on the edge of the steep rock to allow us to set up the screened bug shelter. With no soft earth to peg in either tent, we secured the corners with large stones.

During the early morning hours, with the approaching sound of a thunderstorm, we were awakened. Even though it was dark, we quickly decided to pack our tent and gear in our waterproof bags to stay dry and wait out the storm in the bug shelter. We prepared breakfast and sat in our camp chairs. We enjoyed our morning coffee while taking in the thunder and lightning show. The skies opened up with rain, and the thunderstorm rumbled among the islands. After less than an hour it passed and the rain ceased a short time after.

We paddled through Whalesback Channel where we marvelled at the granite hills, deep blue water and panoramic views. After the early morning thunderstorm and the brief calm that followed, once again, we found ourselves paddling into a headwind all day.

Glenn Reflects:

I was curious about the opening, Little Detroit Cut, a blind dogleg passage carved between the mainland and Aird Island. This opening, with a width of fifteen metres, is the narrowest point of navigable water in the North Channel. What worried me was once we left the protection of Aird Island, the wind and waves might be stronger beyond Little Detroit Cut. I was also worried about encountering a power craft entering the passage from the opposite end. The blind corners and sharp turns gave little time to avoid the path of an approaching vessel. It turns out I worry too much. We did not encounter any boats, and the wind and waves beyond the passage were no worse than what we had paddled on during our time on Lake Superior. We exited the Little Detroit Cut and paddled into a headwind. There is nothing that makes me more tired than having the wind blow into my face all day. Later that afternoon, we came across a large group of young kayakers from a girls' camp heading towards us. Some were tired and expressed their frustration by shouting to us, "It must be nice to have the wind at your back—we have been paddling into the wind all day!" Carol and I thought this was hilariously funny as 'we' thought we had been paddling into the wind all day.

We hoped to reach the Town of Little Current on the northeastern side of Manitoulin Island within the day. Harald, from the group of kayakers at Battle Island that had lent us a marine radio, had been in touch with us. Harald resides on Manitoulin Island with his wife, Laurie. We had planned to camp on the shore but, once again, the kindness of Canadians showed forth when he invited two strangers to stay the night in their home. As we rounded the corner into Goat Island

Channel towards Little Current, we were taken aback by the wind and waves. Watching as powerboat after powerboat rounded the point heading towards Little Current, we headed to the shelter of the shore. The hulls of the powerboats lifted out of the water only to crash down through the waves. Reminding ourselves that the smaller the boat, the bigger the waves, we contemplated how to proceed. After fuelling up with lunch, we dug our paddles into the water and tackled the wind, waves and boat traffic. This combination roughened the waves to an ugly cross-chop that was enough to make one's head spin, making it seem like the motion of riding a mechanical bull. We paddled into the protected waters of the Little Current pier and docked our little canoe among the monsters of the lake. Connecting with Harald and Laurie allowed us to do laundry, take a shower, enjoy a delicious home-cooked meal and sleep in a lovely warm, dry bed. After a hearty breakfast, we headed down to the water. Harald had arranged an interview with the Manitoulin Expositor. The newspaper published the article "Kingston paddlers stop in Little Current as part of three-year Canadian odyssey". Laurie had also posted on the community Facebook 'What's Doin On The Manitoulin' that we would be passing through under the Little Current Swingbridge. It would be a milestone for us as we would be leaving Lake Huron to enter Georgian Bay.

As we waited for the reporter, two kayaks passed by, heading underneath the swing bridge into open water. Once we had completed our engagement with the reporter and had finished saying our goodbyes to Laurie and Harald, we waited patiently in our canoe. Along with larger pleasure craft, together we gathered in the channel and waited for the swing bridge to open. Once the bridge had swung open, we all proceeded forward, one at a time, only to race away into open water beyond. With a final wave from Harald and Laurie, we headed into open water. With an excellent westerly wind, we hoisted the sail and flew down the lake. After many days with the wind in our faces, we were delighted to take advantage of the sail. We quickly overtook the two kayakers travelling along the coastline.

Before we knew it, we were in Georgian Bay. It was exciting paddling along this section of the coast with a multitude of rock gardens; some exposed, most just barely covered by water. The Great Lakes had been experiencing record-breaking water levels this year. The levels were attributed to the tremendous amounts of precipitation and excess runoff during the spring. We had paddled the same area ten years ago and noticed a significant difference on this trip. Where there had been inlets and bays surrounded by whale-back, smooth rock islands, these now appeared to be mostly underwater. Paddling past Killarney Provincial Park, we could see the iconic wilderness landscape which so defines the wild Georgian Bay Coast. We could see La Cloche Mountains' white quartzite ridges lined with pink granite. The terrain comprising La Cloche Mountains are believed to have once been higher than today's Rocky Mountains and were eroded over time to its current altitude. They remain among the highest altitudes in Ontario. Exceptionally clear, sapphire lakes are set among the hills covered with jack pines. Near the fringes of the French River, we travelled around finger points of land until we had reached the interior waters where less wind and waves awaited us. It was the start of a weekend with kayak and canoe trippers out in full force. Official campsites had been occupied early in the day, but we were not concerned as there were hundreds of places to camp on the smooth rock. We paddled slowly, past the windswept pines and rocky islands that dotted the area, totally absorbed with the striking vista before us.

French and Mattawa Rivers
Killarney to Mattawa, Ontario

Dates: August 21 to August 31

Route: Lake Huron (Georgian Bay);
French River; Lake Nipissing; Mattawa River

When planning our route, we had considered continuing further along the coast of Georgian Bay into the Trent-Severn Waterway which would have taken us into Lake Ontario. This route would have been less effort but not as practical for us. It meant we would have had to paddle past our home town of Kingston, Ontario then up the Rideau Canal to Ottawa to finish our cross-Canada trip, only to turn around and paddle back home. Instead, we elected to continue along the historic voyageur routes of the French, Mattawa and Ottawa Rivers. Even though we would have to spend yet more hours portaging—it would be worth it. The French River, officially recognized as Canada's First Heritage River, and historically an important waterway, is a highway of numerous channels and water trails. The scenery along the river amazed us as we watched it transform from the rocky, windswept shores of Georgian Bay to the lush, forested banks of Lake Nipissing. The scenery is pretty with numerous, easy, well-marked portages to circumvent the multitude of waterfalls and rapids.

Along the river, we happened to meet a vibrant couple who had just recently given up whitewater canoeing. Now having decided to paddle strictly flatwater for adventures, they joined us for a conversation about the upcoming river route. They were both in their late seventies, young at heart and, although giving up whitewater canoeing had not been their choice, they said it was the safer route. Due to the decline in their hearing that comes with age, they could no longer hear each other's commands over the noise of the rapids.

Within a few days we reached the entrance of the French River on the West Arm of Lake Nipissing. Thirty kilometres straight across the lake is the city of North Bay. Lake Nipissing is shallow and known to be temperamental and, like Lake Winnipeg, requires constant respect. The lake was to be our last big body of water before we would complete our journey. After we had enjoyed a beautiful campsite on the fringes of French River Provincial Park, we set out early to paddle along the south shore of Lake Nipissing rather than cutting straight across. After we had paddled all day, we reached the South Bay area and decided to spend the night at one of the many sandy coves.

Glenn Reflects:

According to the forecast, the winds were to pick up at noon, so I wanted to head out early, hoping to reach North Bay well before that time. We had just passed the halfway point crossing Callander Bay when I noticed the wind had picked up slightly and a light rain had begun to fall. Our comfortable, relaxed demeanour now switched to hard, measured strokes as we felt an urgency to close the distance and reach the shore. Winds continued to increase. Positioning the canoe to prevent the building waves from striking us broadside, we altered our direction. The rain increased, and the wind, turning waves into large curling whitecaps—the type surfers would appreciate—had become a challenge. We, however, hated this situation. I yelled at Carol above the howl of the wind not to fight the elements but to be one with nature and go with the flow. I am sure she thought I was too severe, but I was nervous and if I appeared calm and confident maybe she would too. The closer we got to shore, the larger the waves became. The driving rain and crashing waves limited our ability to locate landmarks and rocky shoals. After what seemed to be an eternity and, in disbelief that we could be in this situation again, an opportunity appeared. We saw a peninsula which appeared to have a sheltered cove behind it. To reach the cove meant directing the canoe between the point and an exposed rock outcrop. The opening produced a funnel effect for the wind and water to gather, creating still larger waves. Carol was indeed in harmony with nature. She guided the canoe as wave after wave raised the canoe and propelled it forward as if an arrow had been launched from an archer's bow. I was scared stiff and filled with alarm as I felt the stern rise up behind me then accelerate us forward again and again. At times the canoe's stern was so high out of the water that Carol would be stroking into the air. Nervous, but struggling to remain calm, Carol skillfully steered us into the sandy cove where we were able to reach the shore and seek shelter out of the driving rain. What a welcome into North Bay! We learned later that, no matter to whom you speak, everyone has a story about Lake Nippissing.

While in North Bay, we made contact with ottO and returned his marine radio. During a lovely meal that ottO and his partner Jennifer prepared for us, ottO gave us directions on how to portage through the city. The portage was a distance of approximately ten kilometres to arrive at Trout Lake, our next water source.

The following day, after completing the portage through North Bay and paddling to the end of Trout Lake at the mouth of the Little Mattawa River, we established camp at a portage near a dam. After supper, four paddlers approached from the far end of the trail. They were young friends who, after completing a tributary called the Little Mattawa River, were travelling towards the Town of Mattawa. They looked fed up, exhausted and appeared a little cranky with each other. They had left late from Toronto, had spent most of their time driving in heavy traffic and were expecting whitewater river conditions. What they got, instead, were low water conditions requiring many lift overs and the need to drag their canoes, helmets and gear over exposed rocks. They suggested we travel down Turtle Lake and complete the longer portage into Pine Lake to eliminate the route they had just finished. Carol and I took a few moments to assist them with their portage around the dam and helped launch them on their way. We suggested a campsite around the corner. When we went to bed that night, we wondered what lay ahead of us on the Mattawa River.

Mattawa is an Algonquin word meaning "junction of waterways" or "river with walls that echo its current." The river is part of a traditional First Nations route linking the Saint Lawrence River with the upper Great Lakes and was designated a Canadian Heritage River in 1988. We marvelled at the beauty. One could only imagine the early explorers travelling this route, searching for new worlds, seeking out trade routes and opening the way for a nation to be born.

The portages were numerous. In the spring, with higher water levels, we would have been able to paddle or 'line' our canoe through many of the rapids but, at the end of a hot, dry summer, this was not the case. 'Lining' is an age-old alternative to portaging where one uses guide-lines to control the descent of a canoe through moving water. In one instance only, we were able to line the canoe down a rapid. We tied eight-metre ropes to the grab handles at each end of the canoe. From the shore we walked, using the ropes to line or guide the canoe down or up rapids. With years of experience, we have learned how the canoe interacts with the lines and the water. We have to work with the river rather than against it. Struggling against the river may result in the canoe slipping broadside in the primary current and capsizing. Effective communication is as essential when working with a lining partner as it is when interacting with the river.

The British explorer Sir Alexander Mackenzie called the Mattawa "La Petite Rivière." He considered it the most challenging section of the entire trans-continental fur-trade route. We can attest to the difficulty of the portages as Mackenzie did. The Mattawa River drops fifty metres over the fifty-four-kilometre distance to the Ottawa River. Today, the Mattawa is still a busy canoe route where recreational paddlers follow portages unchanged for over three hundred years. Nine of the original eleven La Vase portages are much as the voyageurs found them. All still have their original French names.

There is an annual sixty-four-kilometre 'Mattawa River Canoe Race' from Trout Lake to Mattawa Island Conservation Area. Canoes, kayaks and stand-up paddleboards of all shapes and sizes are placed into different categories and compete for the fastest time to complete racing down the river. The current record to beat is that of 1995: five hours, twenty-seven minutes and fifty-three seconds in the men's division. Regardless of the level, whether pro or children's divisions, the race is a fun event for people who are up for the challenge. We could not imagine how intense this race must be considering the difficulty of the portages. We supposed we could try to speed along to set a personal record; however, we were on our retirement cruise and were enjoying the ride. Nor could we even attempt such a feat with three-carry portages, our bag of wine and our camp chairs in tow. It turns out it took us over three days to paddle the sixty-four kilometres.

We loved the topography of the Canadian Shield. Having spent many backcountry canoe tripping adventures in Algonquin Park which is located just a short distance south of the Mattawa River, we were familiar with this area. During our working careers we were always eager to escape for a few days after a long week of work. On one excursion we headed to an 'unmaintained' area of the park we had always enjoyed. This spot we liked to call the busiest wildlife campsite in the park. Difficult to get to with long, tedious portages but worth the peace and serenity. We have encountered moose, deer and wolves walking through our campsites during the evening hours. One early morning we had a close encounter with a bull moose, an experience we will never forget. It was Thanksgiving weekend in early October. In the predawn hours of Sunday morning, when the forest is still tranquil, nature called. Returning to our tent to get a few more hours of sleep, we buried ourselves back into our winter sleeping bags, zipped up tight and pulled all the drawstrings closed to keep in our body heat. A moment later, a moose made his way towards our site. We could hear the sounds of his heavy footfalls as he slowly made his way to the water. Although we were expecting the moose to walk by, as wildlife has done in the past, he approached our tent.

Glenn Reflects:

I elbowed Carol in a warning. The moose came directly to the tent and stopped. I could feel my anxiety rising and hear my heart pounding in my chest. Carol lying next to me was trembling. Both of us remained motionless and silent. Fearing he would rise and step on our small two-person back-packing tent—with us in it, we did not wish to startle this large-legged beast that weighs close to a tonne. He must have been curious about the object in front of him as he lowered his head and began to sniff the tent right beside my head. I imaged the large, flaring nostrils as he inhaled and exhaled our scent. Not being satisfied, the animal moved to the other side and did the same to Carol. For me, time stood still, and everything narrowed to this moment. Seconds seemed to turn into minutes. We remained perfectly still and mystified. Many thoughts about how this situation was unfolding went through my head. My imagination ran away with me. Even though moose are not inherently aggressive, they will defend themselves if they perceive a threat. What if he did become aggressive? It would be many days if not weeks before anyone would find our crushed and mangled bodies, encased in a thin wrap of nylon that had previously been our tent. The moose returned to my side and began to gently nudge the side of the tent as if he wanted to awaken an object. I thought to myself, does he think we are a female cow lying in wait for a special escapade? As he continued to push against the tent, I resisted the urge to panic. I tried mentally to 'will' him away. When he returned to Carol's side, the nudging continued. I could see the shadow cast from the moonlight reveal a large rack of antlers. Without any response to the attention he was bestowing upon the tent, the beast finally left but not before leaving his deposit of urine to mark his territory. The heavy sound of his footsteps that vibrated on the ground became fainter. We both lay there and breathed a sigh of relief. Our fear was not unfounded as we were well aware that it was moose-breeding season, sometimes referred to as 'the rut'. Moose undergo physiological and behavioural changes. Breeding takes place from mid-September and, for the most part, is complete by the second week of October. We were so unnerved by the event. We thought of every excuse in the book to leave: it was going to rain, it was too cold, it was too windy. We packed up and went home on a beautiful sunny day.

Arriving in Mattawa, we unloaded our canoe at the shoreline of Lynn and Mike's home. Both of us savoured the completion of yet another beautiful Canadian heritage river we had always wanted to paddle. We were treated to a much-appreciated steak dinner, good conversation and the opportunity to sleep in a comfortable bed. After days of portaging between Georgian Bay and the Ottawa River, our tired muscles received a much-needed rest. After a hearty breakfast, we headed down the Ottawa River towards the nation's capital, where we had begun three years ago. The current was in our favour and beaches on the Quebec shoreline were plentiful. We were on the home stretch.

Upper Ottawa River
Mattawa to Ottawa, Ontario

Dates: September 1 to September 11

Route: Ottawa River

Paddling down the Ottawa River offered us a few hurdles: more privately-owned land, several hydro dams and whitewater rapids. The rapids south of the City of Pembroke are renowned for their technical difficulty. They attract thrill-seeking whitewater rafters and kayakers from all corners of the world. For us, however, the rapids were a troublesome barrier. Our capable, but light canoe loaded with gear, could easily be swallowed by the boiling haystacks of churning water. The canoe could be split open on a rock. Also, we did not have the skills to paddle in Class IV rapids. We contacted Jim, an experienced rafter who knows this part of the river well. Jim owns a local rafting adventure company. The plan was to trade in our canoe for a day in exchange for a rubber raft. This way, we would not miss out on this stunning section of the river. Jim worked out the logistics. He agreed to transport our canoe and gear to a predetermined location downstream at the end of the rapids. We joined a group in a rubber raft, and together we experienced the thrill of whitewater rafting. Jim and his team of guides led us on an excursion we will never forget. After a lesson on the water with our new crewmates, we learned how to stroke in unison. We heeded the various commands Jim would shout to us. He skillfully directed the group on how to position the raft to enter each rapid correctly. We spent the day shooting large bus-swallowing waves called the 'Black Hole', 'Butcher Knife' and 'Coliseum' while we were being bucked and tossed about within the confines of a rubber raft. While in calm water paddling toward the next set of rapids, Jim discussed the terrain and the changing landscape the river caused. We arrived at our canoe and set up camp. We felt fortunate that we had paddled this portion of the river under our own steam, just in a different vessel. We had often talked about going on a rafting excursion down these rapids. Now another item has been removed from the bucket list.

A few days after we had left the wild rapids of the river behind, the outskirts of the City of Ottawa lay before us. Not having a place to stay was no worry for us in this big city. Numerous friends and family who live in Ottawa had offered us overnight accommodation. A visit to Uncle John, a member of our SOS Team who conveniently resides on the Ottawa River, was our first stop. As we approached his house, the skies behind us looked threatening and ominous. With binoculars, he watched as we stroked with efficiency and determination to reach shore before the downpour. "Welcome to your bed and breakfast," he offered as we landed onshore. We quickly stored the gear securely in his garage before the rains came—and did it rain! It conveniently stopped the next morning as we headed to Petrie Island on the Ottawa River.

We had discussed with our uncle the navigation through Remic Rapids, an area of low and exposed rocks before we got to the Parliament buildings. He directed us to exit the river at Britannia Bay and walk on the Capital Pathway to the base of the Parliament buildings. The Capital Pathway totals more than six hundred kilometres of paths for cycling, walking, running or rollerblading throughout different green spaces in Ottawa and Gatineau, Quebec. A perfect route to portage a canoe. For the most part, joggers and cyclists did not seem to mind sharing the lane as we pushed the canoe cart along this portion of the Great Trail. We have experienced a significant length of this trail as it connects the nation.

The launch into the river again, just below the Parliament buildings, was timed not to hamper the schedule of the Aqua-Taxi passenger ferry linking Ottawa to Gatineau. Once in the river, we circled in the bay to take photos with the Parliament buildings in the background. We joked that the Prime Minister had forgotten to provide us with a twenty-one-gun salute. Paddling past the opening of the Rideau Canal, we glanced at it, knowing that once we had reached Petrie Island, our cross-Canada journey would be completed, but not yet done. We were going to paddle home to Kingston.

Glenn Reflects:

I am silent, reflecting just how far we have travelled on manual effort—day after day of forward movement, one paddle stroke at a time, one step at a time. I visualize the sights we have experienced, the people we have met, those who have assisted us and the support we have received. In certain ways, I do not wish the paddle to end, yet I am eager to complete the trip and return home. I wonder during the upcoming week or so until we have reached Kingston if our fundraiser will have reached its goal. If not, I speculate we will paddle a few laps around the inner basin of the Rideau River to gain attention, until our goal has been reached.

Our immediate destination now was focused on Petrie Island. Our grandchildren Esther and Rupert, along with their parents, were standing and waving from the shore as they watched us advance. We stopped to pick up Esther and Rupert. We exchanged the kids for a couple of food barrels and positioned the kids in the canoe. The four of us continued to paddle while Esther and Rupert excitedly chatted about their summer.

Petrie Island was where it all had started three years ago. Today was September 11, 2019. The time passed quickly, and we entered into the sandy cove where our journey had begun. We remembered the day as if it were yesterday; how nervous we had been, wondering what lay ahead. We were to realize that slow and steady is a great way to view Canada. One paddle stroke at a time. All with two pieces of wood formed into paddles. From the Pacific to the Atlantic coast was now a memory we would never forget. Tomorrow we would embark on another soon-to-be memory and start the two-hundred-kilometre paddle home to Kingston. Our second stop was just down the road at our other daughter Rachel's home, where a high-fat protein dinner and another good night's sleep awaited us.

Rideau Canal
Ottawa to Kingston, Ontario

Dates: September 12 to September 19

Route: Rideau River; Cataraqui River

We returned to the Rideau Canal, assisted by Robert, our meteo-rologist friend. He had been giving us advanced, long-range weather forecasts since the beginning of our journey. Given predicting weather patterns must be challenging to forecast, Robert's outlooks were rea-sonably accurate. For most of the time, since our trip started in 2017, we had encountered hot and dry weather conditions in all of the Canadian provinces we travelled through. We never experienced days and days of rain that can happen in any part of Canada at any time. The climate and the intensity of the seasons varied significantly all across the country, as we paddled through changing landscapes. Overall, the sun and good weather smiled on us and seemed to follow us wherever we went. While we navigated across Canada, the weather event that had hindered us the most was the perpetual wind. Something that we had not expected to be so constant and forever-challenging.

Robert escorted us with his kayak as we paddled through sec-tions of the canal on which we had frequently speed skated during winter months. A UNESCO World Heritage site, the Rideau Canal Skateway transforms into an epic urban skating rink which winds its way through downtown Ottawa. Stretching almost eight kilometres the Skateway's skating surface is immense, making it the largest natu-rally frozen skating rink in the world. The Skateway is equivalent to over ninety Olympic-sized hockey rinks!

199

Before the locks had closed for the day, we said our farewells to Robert as he paddled back the way he had come. The time we spent returning to Kingston down the Rideau Canal was relaxing. We felt gratified that we had accomplished what we had set out to do. We just had to enjoy the paddle home. The lock staff who controlled the boat traffic were most accommodating, locking us through even if we were the only boat in the lock. Unlike at the Lachine Locks in Montreal, there were camping spots at the majority of the Rideau Canal's lock stations. We often commented that if a family wanted to transition from car camping to backcountry canoe camping, the Rideau Canal would be an excellent introduction. Campsites are available for tenting, hot and cold running water is accessible, and restrooms make it all comfy.

The Rideau Canal was built between 1826 and 1832 and is the best-preserved, fully-working example from North America's significant canal-building era. Its construction through more than two hundred kilometres of bushland, swamps, and lakes was an enormous feat. Each year, as many as five thousand workers, mainly Irish immigrants and French Canadians, toiled under the supervision of Colonel John By, civil contractors and the Royal Engineers. Working in challenging conditions, they endured injury and disease. Hundreds of the workers died. The Rideau Canal chiefly served as a critical artery for moving goods and people until the 1850s. The Canal became a popular recreational destination in the twentieth century and is a monumental engineering feat so well built that, today, much of it remains as it was. Colonel By should have come home a hero. Returning to England in 1832, despite his superb achievement, he was criticized by the Treasury Board for some allegedly unauthorized expenditures. He died a broken man three years later. Imagine if he could see his accomplishment now through the eyes of our world as it is today?

Taking our time, enjoying the temperate fall weather, we arrived at the last lock at Kingston Mills eight days later. The lockmaster was aware of the near completion of our trip. He stood upon the large wooden gates with an assembly of his staff. We began our descent, sinking lower to reach the water below. Once the large wooden doors opened to reveal the passage before us, the staff gave us boisterous applause and offered their congratulations. We felt an emotion of celebration but only for a brief moment, as the lock operator interjected, "You only have seven more kilometres to go...do not screw up." He was right. Now was not the time to become complacent. It was a beautiful day; the wind was calm and we were advancing more quickly than anticipated. With less than a kilometre to go, we were early for the prearranged meeting time of one o'clock that Loving Spoonful had organized for our homecoming. To pass the time, we hid behind Belle Island on the Cataraqui River to wait for the appropriate moment to make our arrival.

Not minutes later, a canoe rounded the island. It was an older Sportspal® Canoe of marine aluminium construction with flotation foam on either side making it very stable. Aboard were two occupants. In the front was a lady with a big-brimmed hat shielding her face from the sun. She was sitting in a lawn chair. Behind her was a rugged outdoor guy paddling the canoe. We were surprised to see Carol's eighty-six-year-old mother and Carol's brother Paul come out to accompany us on the final stretch home. This Sportspal® was the canoe in which Carol had learned to paddle many years ago as a child. It was nostalgic for Carol to see it now at the end of our journey.

With emotions running high for both of us, we were delighted to have reached our destination. Together with other paddlers from Kingston who had come to greet us, we stroked up to the Cataraqui Canoe Club dock—our final destination. Loving Spoonful, staff, supporters, friends and neighbours were all there to welcome us home. Loving Spoonful had organized a grand party complete with local media. What a sweet moment it was.

Epilogue

Several months after we had completed our cross-Canada 'retirement cruise', Canada, along with the rest of the world, was gripped in the global COVID-19 pandemic. It was a difficult time of uncertainty and fear. Many were disconnected from their families and friends. By choice, in our wilderness trips, we seek remote areas distant from civilization. We are familiar with isolation when paddling hour upon hour, day after day or, when windbound, spending multiple days in one spot. We learned to make good use of our time. Making the best use of our time during the pandemic was no different. Like many others, we faced long days of isolation. Life as we knew it came to a standstill. To keep busy, we came up with a plan to write 'Canoe for Change'. Our thoughts were still very much alive with memories of the experience which we wanted to share. When not writing, we were practising 'physical distance' hiking or paddling in the wilds of our home province of Ontario. Some were adventures, while some, misadventures—but that is another book perhaps.

Together with the support of people across the country and at home, our cross-Canada voyage 'Canoe for Change' raised tens of thousands of dollars for Loving Spoonful! Each and every dollar donated is a testament to the fact that fellow citizens believe in working towards a more food-secure Canada. Work continues even more so now that Canada, along with the rest of the world, could face higher food insecurity due to the effects of the global pandemic. Loving Spoonful is working hard with farmers towards establishing a reliable and sustainable fresh food system, not only locally but at a national level.

As if it were a dream, every now and then we ask each other, "Did we just really paddle across Canada?" No matter if you are twenty or sixty, it is incredible what a person can accomplish once they put their mind to it. With sheer determination and a positive attitude, one can achieve almost anything one sets out to do. In the case of paddling across Canada, it was not so much the physical strength that enabled us to stroke over eighty-five hundred kilometres. A significant part is due to attitude. This approach, and being passionate about what we are doing, contributed to the success of our journey: moving along the shoreline at a slow pace, opening our senses to the beauty of the world around us. If we had not had an appreciation for nature, it would have been more of an ordeal rather than a journey. Driven by the joys and excitement of not knowing what lay ahead, we went to bed each night excited to get up the next day. Each year that we embarked on the trip, it took a few weeks to become adjusted to the rigours of the journey. However, once we settled into the routine, a feeling of freedom overtakes—a sense of freedom which feeds our spirit and soul.

In today's times we live in a hectic, sometimes disconnected, world, in a rat race of technology, in fast-moving cities and at a fast pace of life. Living close to the earth, as we did throughout our cross-Canada canoe trip, made us realize that nature is part of us, that it gave back something very special to us. We felt a sense of freedom, purification and an appreciation for our free wild country. We learned along our trip that we were inescapably connecting with nature, recharging from the busyness of the lives we had led for so many years. We found not only a connection to the land but to its people as well. Those we met along our journey instinctively wanted to know more about the connection to nature that they sensed in us. Canadians from coast to coast became part of our odyssey. We were overwhelmed by their acts of kindness and outpouring of support. Perhaps, in their own way, they were trying to grasp this nature within themselves—giving back enabled them to become part of our journey.

Together, we found the most rewarding experience from our coast-to-coast paddle came from the voyage itself, not necessarily the destination. This voyage was like 'life' in a certain way. It started with an idea, it grew within us and then we made it a reality. Once the voyage had begun, we felt committed and, at times, were challenged to the limit. Politely some people would ask, not in these exact words mind you, but, 'How can you stand to be in a canoe with each other day after day?' Carol's reply would be, "We treat each other better than we do our best friend—with love and respect. If we have an issue, we address it right away. We come across many challenging moments dealing with the elements and the uncertainty of the journey ahead. We must work as a team. We either swim together or sink alone. If all else fails, Glenn stays in his corner of the canoe, and I remain in mine." Each new experience made us grow. In life, it tests, you learn, and you mature from experience no matter how old you are. What we had gotten ourselves into was a once-in-a-lifetime journey—a gift we gave to ourselves. Basking in the vastness and beauty of this country we call home we propelled ourselves forward together, unconstrained, one paddle stroke at a time. Enjoying life and taking each moment in stride. We have memories we will cherish for the rest of our lives.

Yes, it took us longer to complete our cross-Canada paddle than expected. Still, it did not matter because we were not out to break any records of the fastest time or go the farthest distance. We did what we loved in the country that we loved with the person whom we loved.

It was a retirement cruise to remember.

Acknowledgements

As we look back on our cross-Canada canoe odyssey, one of the most rewarding memories we have is our interaction with our fellow Canadians. From the Pacific Ocean to the Atlantic Ocean, we experienced boundless generosity and kindness. To all those we met, thank you for allowing us to camp on your front yard, private land, beach, wharf, Pow Wow ceremonial grounds or sacred pilgrimage sites. Thank you for letting us stay in your spare room, your trailer, cabin or gazebo. Thank you for giving us a warm meal and a bed to sleep in; for allowing us to do our laundry or take a much-needed shower. Thank you for allowing us to relax in your sauna! Thank you for taking us into your homes, trusting us, giving us shelter, giving us encouragement and sharing our common bond of being Canadian. Thank you for chasing us down, whether in your boat or in your car, to provide us with food, fresh fish, home-baked bread, fruit or vegetables. Thank you for sending, receiving and holding our prepackaged, dehydrated food supplies. Thank you for waving at us when we were walking down the highway with our canoe, honking your encouragement, giving us directions and providing weather reports. Thank you for stopping to chat with us, cheering us on and giving us hope. Thank you for helping to carry our gear over a portage, to the shoreline or to a campsite. Thank you for being interested in our trip or being concerned for our safety and our well-being. Your offers of assistance, loaning your weather radio or connecting us with others was invaluable. Thank you for fixing, mending or repairing our equipment. Thank you for taking us sightseeing, telling us about your community's history and for sharing your own unique stories. Thank you for giving us an early morning coffee, hot water, drinking water, wine or a cold beer! Thank you for waiving parking, mooring, or campground fees. Thank you for driving us or lending us your vehicle to get supplies! Thank you for assisting us in times of emergency and getting us back on track. Thank you for slowing down in your motorboat or vehicle when passing us. Thank you to the Canadian Coast Guard for steering us out of the shipping lanes and to government personnel for getting us over, through, and around borders, dams and canal lock systems. Thank you for giving us cards, sending emails, following us on social media or our website blog—your words of encouragement kept us going! Finally, thank you for asking about our cause for food security, sharing your ideas for a food-secure Canada and for your generosity in donating! We are eternally grateful to each and every one of you.

We did not have a support vehicle following us as we paddled and portaged across Canada. However, we had a fabulous home support team! Without them, our job would have certainly been more of a challenge than it already was. Thank you to our Uncle John Mowle for providing us with statistics, point-of-interest information, detailed maps and logistics on upcoming hazards. The reassurance of being prepared for the route ahead kept us safe and gave us peace of mind. Thank you to our daughter Rachel Vandertol, our home office volunteer. She assisted with website editing, weekly website blogging and social media troubleshooting. Sorting through hundreds of texts and pictures took patience and dedication which Rachel excels at! Thank you to our brother Paul Vanden Engel for his diligence in tracking us and being on call, day or night, to rescue us if needed. Many hours were spent by Paul and his wife Denise to pick us up after the completion of each leg of our three-year journey. Each of these incredible people spent hours upon hours of time and dedication. They supported us in their own unique way. For this, we will be forever grateful!

Thank you to our one and only sponsor, Dave and Debbie Fitzerman of DFC International Computing, for creating and maintaining our website. The website was an invaluable resource that gave an insight into our journey and, most importantly, our cause!

Acknowledgements

A very special thank you to Loving Spoonful for believing that we could actually paddle across Canada! Thank you to Mara Shaw, a dynamic role model whom we admire and respect for the work she has accomplished for food security in her community. Thank you to Melanie Redman who set up creative media press coverage, collected funds and threw more homecoming parties than anticipated. Thank you for encouraging us, believing in us and allowing us to represent our combined passion for food-secure communities. Being able to access affordable, healthy, fresh food is a right that each and every Canadian should have. That right is why Loving Spoonful works so diligently across Kingston to connect people with good food! Kingston is so very proud of you, and so are we...

We are eternally thankful for our close-knit community of kind-spirited neighbours who rallied behind us, encouraged us, assisted us and cheered us on. Thank you for looking after our home, vehicle and plants while we were away. Thank you to Brian and Ada Pointon and Niall Kenney for always being quick to help out at a moment's notice.

Thank you to our families, for your understanding of missed summer get-togethers and special occasions. Thank you to Debra Laroque-Kovacs for coming up with the name 'Canoe for Change'. What vision! We further thank Paul and Bettina Stanulis who shipped our dehydrated food along our route. The logistics were at times impossible, but you always ensured we received our shipments on time. Thank you to Tony Napolitano and acquaintances for storing our canoe and equipment in Thunder Bay. What would we have done without you!

The publishing journey was a new and exciting experience which, much like our trip across Canada, we had to learn as we went. Thank you to Kayla Lang, our FriesenPress Publishing Specialist for making our transition smooth. Thank you to Susana Giugovaz and Robert Boggs for reading our draft manuscript and for giving us invaluable information. Your honesty is appreciated. We are deeply indebted to our editor Helen Booth. She dedicated many hours of her time and gave us her expert guidance, where it was needed. We will be forever thankful for your kind and generous gift.

We would further like to acknowledge in no particular order: Tom Farr, Larry and Linda Wood, Greg and Triena Partridge, George Bissonnette, Sharleen Martell, Esther Stephens, Rupert Stephens, Jean Pendziwol, Craig McDonald, Gregg Prisby, Juniper Silverthorne, Lorraine and Ralph Will, Jim Coffey, Eric and Rebecca Noort, Peter Quaiattini, Nadine Johnson, Kevin and Kole Dalgleish, Bill and Dianne Fry, Karen and Lee Bowman, Marion Mowle, Daniel Gallant, Sandra Biddington, Catherine and Doug McGregor, Emma Allgood, Patrick Buchanan, Harald and Laurie Simon, John and Cheryl Vandermeer, Jake Vandermeer, ottO Bédard, Jennifer Upton, Denise Vanden Engel, Mike Westley, Lynn Turcotte, Paul Vandertol, Tony and Karen Sepers, Ed, Kevin, Bob, Roland, Lundy, and Steel River Jim. We met and engaged with so many people—all with great stories. Forgive us for not including everyone in our book, but you are in our hearts and we acknowledge you also.

Glenn Reflects: I am extremely grateful for my wife, Carol, for putting up with me. Her determination, skill and perseverance to complete the journey made me realize that enjoyment in life is what one simply makes of it. Without her companionship and compassion, I am sure I would have quit long before completion. Love you.

Eric after hearing your life story and witnessing your gentle nature and positive outlook on the world made us realize how important it is to bring awareness to food insecurity in Canada. You are not alone in your struggles. Godspeed in all your travels—we hope your dreams come true—wherever you may be...

References

United Nations Educational, Scientific and Cultural Organization (UNESCO): Biosphere reserves in Europe & North America, https://en.unesco.org/biosphere/eu-na

Paddling.com: Understanding Tides, https://paddling.com/learn/understanding-tides/

Confederation Bridge, https://www.confederationbridge.com/site/about

Bay of Fundy.com: Bay of Fundy Tides: The Highest Tides in the World, https://www.bayoffundy.com/about/highest-tides/

Government of Canada: St. Peters Canal National Historic Site, https://www.pc.gc.ca/en/lhn-nhs/ns/stpeters

Council of Nova Scotia Archives: Beaton Institute, Cape Breton University, https://novascotia.ca/archives/communityalbums/CapeBreton/archives.asp?ID=698

Canadian Geographic: Fire & Bikes: British Columbia's Kettle Valley,

https://www.canadiangeographic.ca/article/fire-bikes-british-columbias-kettle-valley

BCTrails.com: KVR (Kettle Valley Rail), https://bcrailtrails.com/

The Canadian Encyclopedia: Kootenay Lake, https://www.thecanadianencyclopedia.ca/en/article/kootenay-lake

City of Lethbridge: Parks and Trails, https://www.lethbridge.ca/Things-To-Do/Nature-Centre/Pages/Things-To-See.aspx

Government of Canada: Canadian Forces Base Suffield National Wildlife Area, https://www.canada.ca/en/environment-climate-change/services/national-wildlife-areas/locations/canadian-forces-base-suffield.html

Canadian Feed the Children: Improving First Nations Nutrition, https://canadianfeedthechildren.ca/where/canada/

Glenn Sigurdson: Vikings on a Prairie Ocean, http://prairieocean.ca/the-lake-and-its-people/

The Canadian Encyclopedia: Lake Winnipeg, https://www.thecanadianencyclopedia.ca/en/article/lake-winnipeg

The Canadian Encyclopedia: Lake Superior, https://www.thecanadianencyclopedia.ca/en/article/lake-superior

Government of Ontario: Quetico Provincial Park Management Plan (Published 2018), https://www.ontario.ca/page/quetico-provincial-park-management-plan-published-2018

Government of Canada: Parks Canada: Pukaskwa National Park, https://www.pc.gc.ca/en/pn-np/on/pukaskwa/securite-safety/eau-water

Lighthouse Friends: Battle Island Lighthouse, https://www.lighthousefriends.com/light.asp?ID=1562

Nasa Science: Perseids, https://solarsystem.nasa.gov/asteroids-comets-and-meteors/meteors-and-meteorites/perseids/in-depth

The Canadian Encyclopedia: Mattawa River, https://www.thecanadianencyclopedia.ca/en/article/mattawa-river

Mattawa River Canoe Race, https://www.mattawarivercanoerace.ca/

Ottawa Tourism, Rideau Canal Skateway, https://ottawatourism.ca/en/see-and-do/rideau-canal-skateway

References

Dawn Dickinson and Dennis Baresco: Prairie River, (The Federation of Alberta Naturalists,2003)

Lee Allen Peterson: Peterson Field Guides: Edible Wild Plants, (Houghton Mifflin,1977)

Eric Morse: Fur Trades Routes of Canada; Then and Now (University of Toronto Press,1969).

About the Authors

Carol VandenEngel and Glenn Green have always had a passion for the great outdoors. These passions take them hiking and biking through autumn colours, enjoying wilderness camping, swimming with the fishes, meeting diverse and friendly people, snowshoeing through the woods on chilly winter days, watching the early morning mist on a quiet lake with a cup of coffee, or simply sitting on a patio listening to undiscovered Canadian musicians. As they paddle, they are raising funds and awareness for a local hometown charity, Loving Spoonful, that promotes food security.

After enjoying fulfilling office careers, now in retirement they continue to do the things they love: travelling off the beaten path, maintaining a healthy lifestyle, and always looking for the next challenge.

In her spare time, Carol volunteers to advocate for people with mental health issues, for those who live in poverty, and for causes that support these issues in our society. Glenn has volunteered as a youth softball coach, Scouts Canada leader and currently enjoys participating with several outdoor groups.

Carol and Glenn live together in Kingston, Ontario. Between the two of them, they have four grown children and four grandchildren.

View the authors' website of their epic canoe trip across Canada, their blog and links to social media platforms:

⊕ www.CanoeforChange.ca

✉ info@canoeforchange.ca

Follow 📘 📷 🐦

If you are inspired, we invite you to support the ongoing work of Loving Spoonful by donating to:

Loving Spoonful
263 Weller Ave #4
Kingston, ON K7K 2V4
www.lovingspoonful.org
info@lovingspoonful.org

A portion of the proceeds from the sale of 'Canoe for Change' will be donated to Loving Spoonful.

Printed in Canada